automation in developing countries

Round-Table Discussion on the Manpower Problems Associated with the Introduction of Automation and Advanced Technology in Developing Countries
(Geneva, 1-3 July 1970)

International Labour Office Geneva

ISBN 92-2-100158-X

First published 1972
Second impression 1974

ILO publications can be obtained through major booksellers or ILO local offices in many countries, or direct from ILO Publications, International Labour Office, CH-1211 Geneva 22, Switzerland. A catalogue or list of new publications will be sent free of charge from the above address.

Printed by La Concorde, Lausanne, Switzerland

FOREWORD

Higher employment levels and social progress have once again become central issues in economic development. More jobs and a more equitable distribution of income and status are no longer seen as incidental and often delayed benefits of economic growth, but as prerequisites. Unemployment, frustration, despair and unrest interfere with delicate organising tasks, with capital accumulation, and even with receptiveness to new knowledge. Without the proper setting for organising, saving and learning, productivity will grow slowly, if at all.

This change in perspective poses the greatest problem for less developed countries when they face the choice of adopting or rejecting automation, computers and the most highly advanced technology. Must they forgo its prodigious output-raising but labour-replacing capacity? Must a widening gap in knowledge and experience be accepted? Or can one adopt the new technology selectively, without a disproportionate loss of jobs, harsh training problems, rising inequality, and conflict?

During the early 1960s the International Labour Office began to study the manpower adjustment problems and programmes that followed the coming of automation and electronic data processing in industrialised countries. At that time only a few organisations in developing countries had highly advanced technology, and the occasional computer was used for research, not production. By the end of the decade, however, automated refineries, breweries and chemical plants had been set up in the tropics, and computers had replaced clerks in banks, insurance companies, airlines, public utilities and even cotton mills, not to mention government ministries. French-speaking Africa had an estimated 100 computers; India, the Republic of Korea, Singapore and Taiwan had 160 between them. The Colombian Government alone had 14. These are small numbers compared with the 5,700 computers in Japan (in 1969) and the 60,000 in the United

States, but they seemed to presage greater, perhaps ill-boding, trends. The ILO therefore undertook to study the spread of highly advanced technology in developing countries, particularly its effects on manpower. It is from that effort that this volume has been prepared.

The book opens with seven general and comparative analyses. Professor Nicholas Kaldor fits the phenomenon of advanced technology into a general strategy of economic development. A comparison of the reaction to advanced technology in underdeveloped and developed countries, both in the present and in earlier centuries, is made by Professor W. Paul Strassmann. The experience of Japan merits special attention and is analysed by Professor Gustav Ranis. To highlight the contrasts between Japanese and Indian patterns of adaptation, Dr. Jack Baranson examines the case of diesel engine manufacturing. A broad analysis of organisational requirements, backed by further case studies, is provided by Professor Peter Kilby. The potential and limitations of more effective management for better use of automation in India is shown, with three case studies, by Professor Ishwar Dayal. Finally, Professor Everett Kassalow compares trade unionism in different developing areas, particularly the response to factory and office automation.

To obtain a more detailed view of the issues, the book also presents six case studies. Two of these were drawn from Asia, two from Africa, and two from Latin America. Each of these pairs consists of an industrial case of highly advanced technology in one country and of a computerised service industry in another. Each author has a full understanding of the manpower issues which developed as the new technology was introduced in factory and office. The authors also relate the enterprise to the industrial branch and the branch to the national economy. These case studies are drawn together by a brief section that calls attention to their common and contrasting features, and notes a few of the implications.

During the first three days of July 1970 the authors of both the general and comparative analyses and the case studies came together in Geneva for a round-table discussion. This was organised jointly by Professor W. Paul Strassmann (on leave from Michigan State University) and Mr. Ralph H. Bergmann, who was at the time responsible for the ILO's activities in the field of automation; they also subsequently undertook the technical editing of the various papers for this volume. Dr. Guy Routh, who had collaborated with Mr. Omari S. Juma on the Tanzanian case study, chaired the meeting. Since the papers had been circulated well in advance, the sessions opened with critical analyses of each paper by two or three assigned discussants. General discussion and replies by the authors followed.

The last section of this book, "Summary and Policy Recommendations", is primarily based on these three days of discussion. For the sake of clarity, participants' statements are not given in chronological disorder. Instead, the summary proceeds topically, giving contrasting positions but generally avoiding the tedium of indirect quotations. Direct quotations given in the summary are of words spoken and recorded at the round table. The editors did not, however, freeze the development of ideas on 3 July 1970, but kept the range of included possibilities open until the book went to press.

Only the policy implications were not amended after 3 July 1970. Hence, none are included exclusively on the basis of the editors' opinions. All were advanced before the assembled group and had a chance for critical inspection. This is not to say that these implications are "official recommendations". No votes were taken, and indeed the round table never aimed at such results. Nevertheless, a hope for better policies, where needed, lay behind the entire project. Whether proposals are needed and workable or not must be judged by the logic and evidence presented in the book (or available elsewhere), not by the degree of consensus. To underscore the importance of policy, the major policy implications are listed here.

1. Since automation often means a drastic substitution of capital for labour in response to a minor rise in labour costs or fall in capital costs, it is important that these rises and falls reflect the scarcities of factors of production. Computers and automated machines must not be subsidised directly or indirectly; wages of employed workers must not rise without regard for the chances of the jobless or the underemployed, both urban and rural. Computers and automated machines should not be subject to lower tariffs, higher tax exemptions, easier credit and subsidised servicing. Through participation in formulating a nation-wide wage policy and other measures, trade unions should share economy-wide responsibility for employment. Given trends in population and labour force growth, it is even better to err in the direction of prolonging labour intensity than to adopt capital-intensive methods prematurely.

2. Automation and computers should not be introduced as a means of overcoming shortages in managerial talent. The answer is not prohibition but an expanded management development programme that stresses better supervision, production scheduling, quality control, and plant or office layouts with ordinary mechanisation. Moreover, managers learn informally where they can experiment for gain but must bear the risk of loss. Government policy should not get in the way of such managerial development.

3. Where automated production equipment is priced in accordance with the scarcity of capital and foreign exchange but still produces at lower cost and higher quality, it should be accepted as the best way of giving domestic and foreign buyers the most for their money. Such competitiveness reduces

pressure on the balance of payments and may encourage saving and other growth-stimulating behaviour.

4. Compared with electronically controlled production equipment, it appears that computers in offices are less well used, partly because their introduction is less well planned. Their use is most apt where management can benefit from a continuous flow integrated information system. Before ordering, the firm must know if a good chance for such use exists. This knowledge is not provided entirely objectively by computer salesmen. The firm must know of alternative possibilities for brief rentals of computer time or itself rent out its computer, rather than use it for submarginal, labour-displacing, routine internal data processing. Training and energetic co-ordination policies for computer sharing are needed.

5. Where displacement of workers due to the introduction of automation or computers is indicated, a national manpower adjustment policy is needed to ease the consequent dislocations in a fair and equitable manner. The ILO can assist in the development of such a policy through the formulation of a code or checklist concerning the responsibilities of management, workers, unions and government for coping with the effects of technological change.

6. Further research, particularly the analysis of successful practice, will improve execution of the five policy suggestions that have been made. It is also likely that research can be fruitful in: (*a*) analysing changes in skill and job content as a result of automation in poor countries; (*b*) exploring the psychological and sociological effect of office automation—a new phenomenon; (*c*) examining the adequacy of training programmes and the possibilities for job redesign, especially for supervisors; (*d*) reviewing successful cases of redesigning products and technology for the conditions of developing countries. Such studies should focus on a single industrial branch and, where appropriate, can be carried out by the ILO alone or jointly with other United Nations agencies.

The contributions to this book would suggest that such research will not have to begin from scratch. A good deal of information, insight and experience already exists. Nevertheless, there is a long way to go, and the need for solutions is urgent.

CONTENTS

CASE STUDIES

GENERAL AND COMPARATIVE ANALYSES

ADVANCED TECHNOLOGY IN A STRATEGY OF DEVELOPMENT [1]

NICHOLAS KALDOR
University of Cambridge

There has been a great deal of discussion in economic literature in recent years as to the most appropriate technology that underdeveloped countries should adopt in order to secure the best means of economic development—that is, in order to attain the fastest sustainable increase in real income per head. This problem has become particularly acute in the light of recent advances in the field of automation, electronic data processing and control, and so on.

Two main aspects of this discussion can be distinguished: a wider one, and a narrower one.

The first aspect relates to the strategy of economic development. Given the limitation of available resources, how should they be applied between the different economic sectors? Should they be concentrated on developing technological standards in the traditional economic activities (which means improving the efficiency of agriculture) or on developing new activities, that is to say, on establishing modern industries?

The second aspect relates to the question: What are the most appropriate techniques for adoption in any one industry? Should the underdeveloped countries copy the latest, electronically controlled technologies of the advanced countries; or import something a little older; or should they develop new technologies, appropriate to their circumstances? Should they aim at creating the maximum increase in new employment opportunities, or should they aim at higher productivity per worker, even if this involves a slower increase in the volume of employment?

Neither question permits a simple, general answer; the most that one can expect from economic analysis is for it to set out the relevant factors and establish certain general criteria for judgement. The present paper aims at presenting a brief survey of both aspects.

---

[1] An earlier version of this paper was presented to the Conference on Technological Change and Human Development, Jerusalem, 1969.

THE STRATEGY OF DEVELOPMENT

The postwar development plans of many underdeveloped countries have frequently been criticised on the grounds that they concentrated on industrial development on a broad front, to the neglect of agricultural development; and that they concentrated on developing substitutes for imports, even when this involved a very wasteful use of resources, to the neglect of developing exports. These inward-looking strategies of development meant that, despite very large investments of capital (financed by external aid or through varying forms of taxation of the agricultural sector), they brought a very meagre social return, either in terms of additional employment or in terms of improvement of living standards.

As a criticism of the postwar economic policies of many underdeveloped countries this is no doubt well founded. No one who examines the disappointing record of a number of underdeveloped countries in Latin America or in south-east Asia could fail to be impressed by the fact that the pursuit of indiscriminate import substitution, without regard to cost, led to perpetual balance-of-payments problems, inflation, low rates of growth, and meagre results in terms of additional employment or higher consumption per head of population. Yet it is a criticism concerning methods, rather than aims; of tactics, rather than of long-run strategy.

For there can be little doubt that the kind of economic growth which involves the spread of modern technology, and which eventuates in high real income per head, is inconceivable without industrialisation. In that broad sense there are no alternative roads to economic development. It is no accident that all advanced countries with relatively high incomes per head have a large manufacturing industry and, in most cases, are also large exporters of manufactured goods.

The reason for this is not only that, as real income rises, a diminishing proportion of income is spent on food and a growing proportion is spent on industrial goods and on services. One could, in theory, conceive of a country specialising entirely in agriculture and obtaining all its industrial requirements from abroad. But it could never become a high-income country simply because technologically developed agriculture could never absorb more than a fraction of the working population on the available land. Though in all underdeveloped countries the greater part of the working population is "occupied" in agriculture, most of this represents disguised unemployment; a rural community maintains all its members and expects everyone to share in the work. Much the greater part of this labour could be withdrawn from agriculture without any adverse effect (and probably with a beneficial effect) on total agricultural output, if alternative employment opportunities were available; for the relief of the pressure of labour on the land is itself a most potent factor in inducing improvements in technology which raise yields per acre, as well as the yield per man. These improvements normally require an increase in the capital employed on the land; but the savings necessary for the increase in capital are themselves a by-product of reduced population pressure. The reduction in the agricultural population and the increased use of capital in agriculture are thus different aspects of the same process. As there are fewer mouths to feed, the "agricultural surplus" rises (the excess of agricultural production over the self-consumption of the farming population). The rise in the surplus enables the farmers to plough back a higher proportion of their output (in the form of better tools, improved seeds, fertilisers, etc.), and such improvements tend to be both "labour saving" and "land saving": they diminish the labour requirements at the same time as they increase the yield of the land.

Hence a technological revolution in agriculture leading to a faster growth in output is generally associated with a steady reduction, not an increase, in the agricultural labour force. The best proof of this is to be found in the fact that those advanced high-income countries which have specialised in exporting agricultural products and importing manufactured goods (and which export the greater part of their agricultural output), such as Australia, Denmark and New Zealand, have nevertheless only a low proportion of their labour force in agriculture—of the order of 10 to 20 per cent, as against the 70 to 80 per cent in low-income countries which barely produce enough to feed their populations on a very low standard. And it is one of the best-established generalisations of economic history that with improvements in technology, and a rise in real income per head, there is a continued reduction in the proportion of the labour force employed in agriculture.

The advance in agricultural technology thus depends on, and is conditioned by, the growth of the agricultural surplus. At the same time, the growth of this surplus is the main factor determining the growth of employment opportunities in the non-agricultural sectors of the economy. This has two aspects.

In the first place, the growth of the non-agricultural employment potential depends on the rate of growth of marketed (as against self-consumed) food supplies. Food is the "wage good" par excellence, and any attempt to increase urban employment at a faster rate than the agricultural surplus permits, is bound, sooner or later, to be vitiated through violent inflation. In the second place, the growth of the agricultural surplus is an essential condition for providing the growth of purchasing power necessary for sustaining the growth of demand for industrial products. The growth of industrial activities is generally regarded as being conditioned by the rate of capital accumulation in industry. No doubt the accumulation of capital is an essential part of this process; in capitalistic economies, however, the process of industrial capital accumulation is very largely self-generated: it is both motivated and financed by the reinvestment of business profits. Hence the rate of capital accumulation is itself conditioned by the growth of profits associated with the process of expansion, and with the growth of profitable investment opportunities provided by the increase of effective demand for industrial products.

Although the expansion of industrial production itself provides an element of this growth of demand, since part of the incomes generated by industrial activities is spent on goods produced by the industrial sector, this self-generated component of demand cannot alone be sufficient to make an increase in production profitable. The growth in demand, which has a determining influence on the pace of expansion—both of the growth of production and employment and of the growth of productive capacity—must be external to the industrial sector: it reflects the increase in the supply of other goods (mainly food and raw materials) for which the products of industry are exchanged. In an advanced economy, with a highly developed manufacturing sector, the most important exogenous factor in the growth of demand is the increase of world demand for its exports. But for a country in the earlier stages of industrialisation which is unable to break into the export markets, the exogenous component of demand is the surplus of its own agricultural sector.

Hence in any sustained process of economic growth, the expansion of the agricultural surplus provides the source of demand for the growth of industry; the growth of industry, with its manifold associated tertiary activities, provides the source for the growth of urban employment; and the latter provides the resources, and the incentives, for technological improvements in agriculture, which in turn, by raising the agricultural surplus, ensure the continued growth of demand for industrial goods. Thus industrialisation and the growth of agricultural productivity go hand in hand, and are complementary to one another. Any improvement in either one of these sectors facilitates the growth of the other, in the manner of a chain reaction.

This has undoubtedly been true of all the present developed countries which have attained their privileged position (with levels of real income per head that are 20 to 30 times as high as in the underdeveloped countries) only as a result of a long-sustained process of rapid technological improvement in both manufacturing and agriculture. In the countries which failed to participate in this process of long-sustained growth, some constraint must have inhibited the chain reaction process: for one reason or another, the environment was not favourable, or not sufficiently favourable, for this process really to start, or if it did start, to be long sustained.

I should like to distinguish between four different factors which could provide the cause of such constraints; this list is not intended to be exhaustive, and in any particular historical case there may have been other (largely political or educational) factors at work.

1. The first of these is the lack of responsiveness in agriculture to outside stimuli on account of economic or social factors. In countries in which agricultural overpopulation has reached very high levels and where, in consequence, the rural community is very poor, the response to better market opportunities may be well-nigh non-existent. Harvests may vary with weather conditions, but a good harvest simply means less starvation: it does not generate a higher cash income, nor provide the opportunities for improving the land through more investment. Even when the physical conditions are more favourable and the population pressure is not so great, ancient forms of land tenure, the survival of feudal institutions, or absentee landlords, for instance, may make it impossible for the actual cultivators of the soil to exploit the opportunities for improvement in the arts of cultivation. It is no accident that institutional reforms in the system of land ownership played such a critical role in the process of industrialisation in Europe. In England, as elsewhere in Western Europe, the so-called "agricultural revolution" historically preceded the Industrial Revolution. In some countries, as in England, this was brought about by the landlords expropriating the hereditary tenants; in others, as in France, by the hereditary tenants expropriating the landlords. Land reform, with the consequent agricultural revolution, also played a vital role in the development of Japan after the Meiji Restoration. By contrast, in countries where this agricultural revolution failed to occur (as in many countries of Latin America, the Middle East and south-east Asia), industrial development was stifled despite State support; economic development failed to reach the stage at which it became self-sustaining through rising levels of real income.

2. A second factor, equally important, is the failure to develop exports *pari passu* with the growth of the economy. For the early industrialisers who were able to supply the world with the products of advanced technology, the rate of industrialisation was export-led from the start. For the late developers this presented a far more difficult problem, since they were trading the advantages of low wages against the superior technology of the older industrial countries; but those which succeeded in passing through the critical phases to become high-income countries were invariably the countries which managed to sustain a buoyant growth of their exports.

For many of the countries outside Europe the initial stimulus to development came from the growth of exports from plantation agriculture and mining. With the growth of industry in Europe, markets were created for temperate foodstuffs, tropical products and minerals. The exploitation of these opportunities proceeded sometimes on the initiative of native producers, but more frequently through European capital and enterprise. It is often contended that the growing foreign-controlled mines and plantations were foreign enclaves which contributed little to the economic growth of the countries in which these developments occurred. This is broadly but not wholly accurate. They brought with them the educational stimulus of foreign contacts; and what is more important, they were a source of export earnings which could be channelled, in a suitable political environment, to provide opportunities for developing local industries. By contrast, the countries which did not have resources for the development of exports of primary products (because of an unfavourable climate or a lack of minerals in their territory) were under the greatest handicap to get economic development started.

3. Export earnings through agriculture or mining do not, however, in themselves suffice to launch the process of industrialisation without State support in the form of subsidies to industry or the adoption of protective tariffs. In the absence of these, the high initial costs (in terms of agricultural products) of home-produced manufactures imposes too severe a handicap on any latecomer to industrialisation to make manufacturing activities commercially profitable. The advantage of any underdeveloped country in the industrial field resides in low wages. In the initial stages of industrialisation this advantage is more than offset by low productivity. Hence under conditions of free trade, when the domestic price of manufacturers in terms of primary products is determined by world prices, the process of industrialisation may never begin.

4. However, whilst the competition of more advanced countries may prevent the domestic establishment of industries in the absence of some measure of protection, such policies are likely to succeed only if they are applied sensibly and with moderation. Excessive or indiscriminate protection may itself inhibit continued development, for a number of reasons. In the first place, any degree of protection involves a levy on agricultural producers, who are forced to sell their produce on less favourable terms in relation to the industrial goods for which they are exchanged; if the protection is excessive, the terms of trade will deteriorate so much as to deprive the farming community of the advantage of a growing real income, which is indispensable for the adoption of more advanced technologies. Hence it makes for agricultural stagnation, which will sooner or later bring industrial expansion to a standstill.

Import duties are efficacious in promoting industrialisation so long as there is scope for creating an internal demand for home-produced manufactured goods through the replacement of the pre-existing imports of such goods. But once the limits of "easy" import substitution have been reached, the momentum for further industrialisation is virtually exhausted—particularly where this development was only brought about by slowing down the growth of agricultural production. For as soon as import substitution is accomplished, the further growth of domestic industry becomes dependent either on the development of industrial exports or on the growth of production in the complementary sector of the economy, that is, in agriculture.

However, industrialisation fostered through high tariffs itself militates against the development of exports. Where the support to industry takes the form of a protective tariff and not of a direct subsidy, the internal price structure is adapted to the internal cost structure—not the internal cost structure to the external price structure. Industries are developed on the basis of a price relationship between manufactured goods and primary products which is divorced from the prevailing world price relationship; and the higher the protection, the greater the deviation between the system of internal prices and world prices will tend to be.

The theory of protective tariffs as a means of industrial development was originally presented as an "infant industry" argument: once the industries are well established, the protection should gradually be withdrawn, or else the industries will fail to become competitive in export markets. But this assumes that the initial degree of protection is none too high, and the degree of protection fairly uniform between different industrial products. This has definitively not been the case in those countries which followed an inward strategy of economic development, that is to say, in those which attempted comprehensive industrialisation on a broad scale, with each separate industry obtaining the differential subsidy required for domestic manufacture, regardless of comparative cost. This meant that the cost of industrialisation was too high, for it involved an excessive burden on the agricultural sector; and since the growth of total employment is always limited by the growth of the non-agricultural employment potential, the growth of industrial activities as a whole was held down. Furthermore, the hoped-for improvements in productivity failed to materialise (or did so only to a moderate extent). With so many industries established more or less simultaneously, none of them could reach a sufficient size to become efficient: the economies of specialisation and large-scale production tended to get lost.

The dangers of such inward strategies are well illustrated by the history of many Latin American countries. For example, Argentina, Brazil and Chile each passed through a phase of relatively rapid growth, following the establishment of highly protective tariffs or import prohibitions during the Great Depression. But in each case this phase was followed by a prolonged period of very slow growth or stagnation, combined with prolonged and violent inflation, as the growth of urban employment tended to rise faster than the growth of marketed food supplies. The conditions under which the initial phase of rapid growth were attained made subsequent stagnation inevitable.

By contrast, the history of many of the smaller European countries which have been comparative latecomers in industrialisation, such as Switzerland or the Scandinavian countries, shows the advantages of an outward strategy. In each case tariffs were kept low and reasonably uniform; this made specialisation in industrial development possible from the start; some industries were developed to the stage of becoming internationally competitive before starting with the establishment of others; the range of domestic production of manufactured goods was broadened only gradually. In this manner the growth of exports kept pace with the growth of production; this provided the means for a steady growth in the employment potential of the industrial sector, and thus made it possible for the disguised unemployment in agriculture to be gradually liquidated.

THE CHOICE OF TECHNOLOGY

I should now like to turn to the second of the two aspects mentioned at the beginning of this paper, which also has been the subject of a great deal of controversy. What kind of technologies should the underdeveloped countries adopt in order to secure the fastest rates of growth?

There has been much criticism of the policies of underdeveloped countries wishing to imitate the advanced technologies of the Western industrial countries without inquiring whether these are appropriate to their circumstances. The advanced countries employ techniques which require a very large amount of capital per worker; since the underdeveloped countries have plenty of labour and very little capital, they should, according to this view, develop intermediate technologies which have a lower capital requirement and a higher labour requirement per unit of output.

Allied to this point of view is the suggestion that, since the great majority of their populations live in rural areas, they should avoid the high economic and social cost of large urban conglomerations and, instead, bring industrialisation to the villages. The development of cottage industries was at one stage an important feature of Indian economic planning; some years later, the idea of creating small-scale metal smelting and engineering works in the villages was the main feature of the "great leap forward" phase of the People's Republic of China. Both of these proved costly failures.

The fact that in all known historical cases the development of manufacturing industries was closely associated with urbanisation must have deep-seated causes which are unlikely to be rendered inoperative by the invention of some new technology. The regional concentration of industries has very important advantages which go well beyond the economies of large-scale operations. They are to be found in the availability of specialised skills, know-how and easy access to markets which make it profitable for firms using similar or related processes to be located close to each other. The concentration of particular phases of cotton textile production in particular cities (and sometimes in special regions of cities) in Lancashire was not the result of any deliberate planning: it emerged spontaneously, as a result of market forces. The same is true of the concentration of the engineering industry in Birmingham and of the jewellery industry in the little town of Pferzheim in Germany (which at one time had 20,000 jewellery workers); one could mention innumerable other examples. In none of these cases could the geographical concentration be explained by the technical need for particularly large-scale plant. The advantages of geographical concentration lay in the opportunities for a higher degree of specialisation between different enterprises and the consequent subdivision of industrial processes; in the availability of labour with many specialised skills; of a wide range of engineering and marketing knowledge, and so on. Though the cheapness of transport and the increasing use of electricity as a source of power made the location of industries far less dependent on their proximity to sources of raw materials and fuel, none of these factors lessened the advantages of a close concentration of industrial activities in large urban centres. Rural industries are unlikely to show the same continued tendency to technological change and improvement which comes from easy communication and shared experience.

In the same way, the idea that underdeveloped countries would stand to gain from the use of special labour-intensive technologies is of doubtful validity. It is of course obvious that countries which are short of capital should use techniques which make the best use of capital, that is, which have a low investment requirement per unit of output. It is a mistake to believe, however, that more primitive or less mechanised techniques which require less capital per worker are also more economical in capital per unit of output. A lower capital/labour ratio does not necessarily imply a lower capital/output ratio—indeed, the reverse is often the case. The countries with the most highly mechanised industries, such as the United States, do not require a higher ratio of capital to output. The capital/output ratio in the United States has been falling over the past 50 years whilst the capital/labour ratio has been steadily rising; and it is lower in the United States today than in the manufacturing industries of many underdeveloped countries. Technological progress in the present century led to a vast increase in the productivity of labour, but this was not accompanied by any associated reduction in the productivity of capital investment.

One reason for this is that the greater speed of "through-put" with modern technical processes led to economies in the amount of working capital required per unit of output. It was found in India, for example, that non-mechanical processes like hand-spinning or weaving are more costly in capital than modern machine processes, despite the fact that the cost of a hand-spinning wheel or a handloom per worker is only a small fraction of the cost of power-driven machinery per worker. The reason for this was found to be the very much greater requirement for working capital in the case of hand-made processes. Just because processing takes so much longer with non-mechanical methods, the working capital locked up per unit of raw material input, or per unit of final output, is much greater.

Nor is it correct to suggest that, just because there is a large amount of surplus labour, labour ought to be regarded as a "free good". However great the level of open or disguised unemployment, employed labour still has to be paid wages. A higher level of employment means a higher wage bill, and a correspondingly higher demand for consumption goods, particularly for food. This is not only because at a low standard of living workers prefer to consume any additional income. Employment requires a greater expenditure of physical energy than idleness; a man can perform a full day's work only if he has a much higher calorie consumption than is necessary to sustain life in the absence of work. For this reason it is often profitable for employers to pay higher wages than the minimum at which they can hire workers: when earnings are low, calorie consumption becomes insufficient for efficient work performance—hence, beyond a point, a lower-wage worker involves a higher wage outlay per unit of output. This, I believe, is the main reason why even in the least developed countries with the largest ratio of surplus labour, the wages of factory workers are so much higher than the "opportunity cost" of labour, that is, the level of earnings per head in the rural sector. Indeed, the difference between industrial wages and agricultural earnings tends to vary in inverse proportion to the degree of development of an economy.

This is not to suggest that the need to maintain efficient work performance is the only, or even the main, reason for the wide and growing earnings differentials between the organised industrial sector and the other sectors of developed countries. Indeed, the need to ensure a certain calorie consumption can hardly explain why in many underdeveloped countries real wages in industry are continuously rising, not only in terms of the goods which the workers produce, but also in terms of the goods which they consume, i.e. in terms of food. This phenomenon has not in my view been satisfactorily explained: it may have something to do with the existence of union bargaining, with legislative controls, and also with the monopolistic character of the manufacturing industries of underdeveloped countries, as a result of which a reduction in costs is not normally passed on in prices, so that, with rising productivity, only rising money wages can ensure that the workers share in the benefits of greater productivity.

But whatever the reason may be, there can be no doubt that the fast rate of increase in wages in the manufacturing sectors of underdeveloped countries (both in money and in real terms, and in relation to the rate of growth of earnings in the economy as a whole) represents a serious handicap to their industrial development and to the growth of employment in the organised sectors. The main reason for this is that, on account of this mechanism, the growth of productivity is prevented from having its normal effect in lowering the prices of industrial goods in terms of agricultural product: it does not therefore lead to any increase in the purchasing power of the rest of the community for industrial goods; nor does it involve any adaptation of the internal cost structure to the world price structure which, as I emphasised, is a precondition for the development of exports.

This process of continuously rising wages may also involve the adoption of a more capital-intensive technology (and therefore a smaller volume of employment for any given output) than would be socially desirable. For reasons mentioned earlier, however, this particular factor may not be of such importance as it is often thought to be in theoretical literature, precisely because technologies involving a lower capital/labour ratio do not necessarily imply a more economic use of capital per unit of output.

All this does not mean that the use of the most highly mechanised technologies, yielding the highest output per worker, is necessarily the best means to be chosen. Nor is it correct to suppose that the best technology for underdeveloped countries is that which yields the highest output per unit of investment (irrespective of labour requirements) as would be the case if labour could really be treated as a "free good". The best general rule, as several economists have pointed out, is to choose the techniques which, at the prevailing and expected level of wages, yield the highest rate of profit per unit of investment: for, as a general rule, the techniques which yield the highest rate of profit serve to maximise the ratio of output to the additional consumption generated in producing that output; the greater the additional output available for reinvestment, the higher the rate of growth of capital accumulation which the economy can sustain.

The most profitable technology for a capitalist employer may thus be no different from the theoretical optimal technique in a planned socialist economy. The question which needs further examination is whether this "most profitable" technique is necessarily different, in underdeveloped countries with relatively low wages per head, from the optimal techniques in the advanced countries with high wages per worker.

If the above characterisation of technological progress is correct, and the latest technologies normally yield the highest output per worker without any greater outlay of capital per unit of output, then the latest techniques must be superior to all others, irrespective of the level of wages, and irrespective of the amount of capital required per worker. However, this may not be true of all cases or of all industries. A highly capital-intensive technology (in the sense of one requiring a large investment per worker) may have a relatively lower labour productivity in a less developed country than in an advanced country as against less advanced techniques. Owing to the greater difficulty of operating complex machines, which are liable to mechanical breakdowns, these may be relatively more costly to use in underdeveloped countries than simpler technologies requiring less skill. A bulldozer, for example, may be the cheapest way of moving earth for road-building or dam-building, in terms of both capital and labour, in a technically advanced country; yet it may be more expensive to operate than simpler instruments in a less advanced country if, in the latter, it is idle half the time owing to mechanical breakdowns, or requires a large force of mechanics to keep it in good repair.

On the other hand, it is also true that in many cases the use of the most advanced technologies saves skilled labour, which is particularly scarce in underdeveloped countries; it may also be contended that in many cases the alternative to the use of such techniques is a more complex human organisation requiring skills of organisation and management which are even more difficult to obtain than capital. Therefore the question whether the most advanced technologies are more or less suitable for underdeveloped countries than for developed countries is one of considerable complexity, the answer to which might well differ in particular cases.

Apart from this, there are two main reasons why the latest machinery may not be the most suitable for use in underdeveloped countries.

The first derives from the opportunity to buy, at a relatively low cost, second-hand machinery which, though physically perfectly fit for use, is no longer profitable to operate in the advanced countries, owing to the competition of more recent and technologically more advanced plant and machinery. It is well known that equipment in advanced industrial countries is withdrawn from use long before it is physically worn out, owing to the continued rise in wages in relation to product prices which causes rapid obsolescence. This gradually eliminates the profit on their operation long before the physical efficiency of the equipment is impaired through wear and tear. Such obsolescent equipment can be acquired at low prices; and in countries where wages are low relative to product prices, they may thus be more profitable to install than the latest new equipment, which is more costly. There is a considerable international trade in such second-hand equipment: and given the opportunity to buy such machinery at low prices relative to their original cost of production, underdeveloped countries may well benefit from adopting, not the latest technologies of the advanced countries, but the technologies that were the most advanced 10 or 15 years ago.

The second reason is that the most advanced technologies require too large an output for their optimal utilisation. Technological advance is biased in favour of large-scale production: the most efficient generating station, the most efficient steel mill, the most efficient tanker, and so on, require a much larger rate of through-put than was the case a generation ago. In 1950 the optimal output of an integrated steel plant was around 2 million tons a year. At present it is around 6 million tons, and in another 10 to 20 years it is likely to be twice as much again. This is partly because there are always inherent economies in large-scale operations; and technical advance consists in overcoming the constructional problems of making things to an ever larger size (for example, the largest tanker that can be built is necessarily the cheapest, since the carrying capacity of a larger tanker increases at a higher rate than the labour and materials cost of its construction, or the labour and fuel cost of its operation). It is partly also because much of the research and development expenditure in advanced countries is undertaken by giant firms whose economic interest is to develop techniques which give a differential advantage to the large-scale producer. The underdeveloped countries whose market for industrial goods is limited have neither the resources nor the market opportunities for the installation of plant with a very large capacity. Perhaps the most important respect in which they require technologies different from those of the advanced countries, and for the sake of which they would be justified in incurring heavy research and development expenditures of their own, is in the development of ways in which modern technological processes in the steel and chemical industries, for instance, could be adapted to efficient operation on a smaller scale.

Finally, one cannot emphasise too strongly that the efficiency and speed with which modern technology can be introduced in underdeveloped countries is very much a matter of the quality and character of the system of education. In many underdeveloped countries too much is spent in providing high-level education of a theoretical kind, and too little in creating a cadre of qualified managers, engineers and technicians. Together with the creation of efficient transport and communication systems, this is the most important aspect in which the development of a social infrastructure promotes economic development.

THE RESPONSE TO AUTOMATION AND ADVANCED TECHNOLOGY: A COMPARISON OF DEVELOPED AND DEVELOPING COUNTRIES

W. PAUL STRASSMANN
Michigan State University

Whether a country is rich or poor, a sharp advance in production methods hits managers and workers first and the general public and government next. How each group is affected and responds depends on the features of the new technique that help or threaten that group's well-being.

Focus on highly advanced technology or use of the label "automation" implies that here is a cluster of production techniques with special features that need attention. The first part of this paper examines whether special features have indeed changed output, material input (including equipment), human input, or all three. The second part takes the responses to these features by management, labour and government in less developed countries and compares them with responses in advanced countries.

SPECIAL FEATURES OF ADVANCED TECHNOLOGY

The most superficial way of defining advanced technology is simply to call it the most capital-intensive. The implication follows that poor countries will not rationally use advanced methods because they save little and should spread their capital thinly. If there is competition for capital, its price will rise compared with wages, making advanced methods costly and unprofitable. If labour-intensive techniques do not exist for a commodity, it should be imported from countries where capital is cheap. Economists long ago learned to modify this simple picture with variations in transportation costs and economies of scale. Every first-year student of economics knows this case and its obverse: why advanced countries rationally use handicraft technology where volume is small and long-distance purchase too inconvenient.

Output changes

If automation and highly advanced technology create special problems and opportunities, more must be involved than these well-known cost patterns. In the first place, technological change need not depend on rising volume and economies of scale. Ultra-modern techniques can reduce costs, including capital costs, without an obligatory expansion of demand and output, and perhaps even with a contraction. As an example, a railway may use a computer to monitor the flow of traffic and can, therefore, reduce its investment in track and rolling stock for a given volume of transport. Indeed, the consolidation of a railway system and a shrinkage of output may be a prerequisite for computerisation, higher efficiency and greater absolute profits.[1]

Similarly, in river basins, computers may facilitate the optimal schedule of water for power and irrigation in accordance with variations in precipitation, soil differences and cropping patterns. Here "the pay-off from application of advanced technology is tremendous".[2] Manufacturing and service industries can vary their input and output mix in response to changing economic conditions through integrated information systems. Where variation is impossible, a sophisticated analysis may be necessary only when the plant is first designed. In all these cases, volume may rise because of greater capacity or lower prices and demand elasticity, but the cost of the capital and labour employed in the analysis could be recovered even without the rise in sales.

Some ultra-modern techniques may actually make sophisticated, capital-intensive production methods viable for production runs that were previously too short. Some complex machine tools could not compete with more rudimentary methods until they were made even more complex and advanced with a numerical control system. The ease of changing programmes makes it possible to shift the machines quickly from one job to another, hence keeping them fully employed and amortised over a larger volume of output. Electronic tape and related hardware make it possible to string a number of short production runs together, simulating a single long run.[3] Devices that allow

---

[1] The writer is indebted to Mr. Karl Gunther, ILO, for this case. See also United Nations, ECAFE, Transport and Communications Committee, Railway Sub-Committee, 10th Session, New Delhi, 13-21 November 1969: *Computers and Cybernetics* (New York, doc. E/CN.11/TRANS/Sub.1/56, Rev. 1; mimeographed), p. 80.

[2] Stephen A. Marglin in Ward Morehouse (ed.): *Science and the Human Condition in India and Pakistan* (New York, Rockefeller University Press, 1968), p. 191. Mansfield reports that the Westinghouse Electric Corporation used a computer to close 6 of its 26 warehouses and to reduce inventories by 35 per cent and yet render better service to customers. See Edwin Mansfield: *The Economics of Technological Change* (New York, Norton, 1968), p. 136.

[3] W. Simon: "Technical and Managerial Innovations Present and Prospective: The Machine-Tool Industry in Europe", in Organisation for Economic Co-operation and Development (OECD), Regional Trade Union Seminar, Paris, 8-11 Oct. 1968: *Education and Training for the Metalworker of 1980: Final Report* (Paris, 1968), p. 32.

quick mould changes perform a similar function in glass, plastic and metal fabricating.[1] Nevertheless, in most cases electronic data processing, including feedback devices, is more valuable if speed is important, if the number of units handled per day is great, and if high precision is needed.

A feedback or computing device, once installed, can often perform at negligible extra cost a great variety of extra services. A product can be refined and tested with greater subtlety. Financial and production records can be amended and analysed in surprising new ways.

The possibilities are too many and varied for a complete catalogue. The social responses to output effects of advanced technology today are, therefore, responses not only to changes in volume and cost, but also to greater variety and higher quality. This feature may be contrasted with the early nineteenth century when technological change lowered cost along with quality and raised volume along with monotony. In that era obsolete methods were less vulnerable than they are in poor countries today. These output and profit effects mean more to government and management than they do to labour, which dreads input effects.

Material input changes

Output characteristics indirectly force social and economic adjustments, but input changes have caused the greater disturbance. One must obviously distinguish between material and human inputs. "Material", in this sense, includes equipment. Changes in all inputs simultaneously are related to a vision of production as an integrated impersonal system that responds smoothly to central control. This vision is the opposite of man's early view of production as a creative act that fuses the worker's personality with the slain deer, threshed corn, carved utensil or woven fabric. According to the modern concept, personality is drained not only from the object but also from man as the producer. Although slaves in mines and galleys suffered this fate in antiquity, depersonalisation for profit, yet without total coercion, followed the factory system. The early factories were indeed modelled on prisons and pauper workhouses, and crude coercion disappeared only with a lag.

Around 1900 attempts were made to integrate workers impersonally into the productive system with the stop-watch and the cash bonus. Such integration did not get very far unless continuous flow devices and conveyor belts helped to dictate the pace. But the vision of impersonal integration did not approach reality until information was generated, transformed, stored, transmitted and applied through non-human agents in the production process.

[1] For examples, see W. Paul Strassmann: *Technological Change and Economic Development* (Ithaca, NY, Cornell University Press, 1968), pp. 170-71.

Mechanisation assigns the work of hands and feet to equipment. Automation adds a capacity to observe and to process information into the operation, usually through electronic means.

Information storage, in the form of loom heddles, templates, cams, discs and even tapes as a means of controlling a mechanism at a later time, has a long history. These devices suggest, if anything, that it must be the effect on the total process that makes automation special. Automation at present exists in advanced form in entire plants in the production of cement, petroleum products, certain chemicals and electric power. Partial introduction has occurred in other utilities, the glass, paper and cigarette manufacturing industries, banking and insurance. These are not exhaustive lists. Practice in poor countries, as we shall discuss under "High-level management" below, has not lagged much behind that of advanced countries in such industries. Old plants are scrapped or overhauled a decade or two later, and new plants may lack the latest novelties; but the difference does not go beyond that. On the other hand, in apparel and construction, for example, where the advanced countries have made relatively less progress, the lag of poor countries is somewhat greater.

The principal material input changes for automation are more sophisticated machinery with more complex control mechanisms, together with punch-cards, tape and other auxiliary supplies. Since newly industrialising nations are least qualified for producing such equipment, some additional pressure on the balance of payments is a likely result of automation. Indirect effects may offset this pressure. Even where foreign branch plants for making computers have been set up, however, local participation and adaptation are avoided and, therefore, "learning by doing" is minimal.

Changes in manpower needs

When a brand-new computer with greater versatility and a superior memory replaces several aged models that cannot learn, adapt or remember well, society considers the switch a business proposition and leaves it to the manager's judgement. When the substitution involves firing aged human inputs who are similarly unadaptable, forgetful or just superfluous, the manager's judgement is open to challenge. Even the economist's bland phrase that "resources are being freed for other industries" becomes suspect. People become so apprehensive that they respond to some consequences that are merely feared as possibilities. These responses will be considered in the next section. Here we shall try to separate actual from the merely possible consequences of automation for human inputs.

A distinction between "mechanisation" and "advanced technology" can be made in terms of the effect on numbers and skills at the plant level. In mechanisation, with a given volume of output, employment or the skill level falls, or both fall. In advanced technology, fewer but more skilled workers supervise and service self-adjusting machinery. But if volume is brought into the discussion, then either type of technological change need not result in lower plant-level employment.

This beginning may be refined with a long list of other distinctions. One may speculate about the effect of advanced technology on the age distribution, sex proportion, education requirements, training programmes, relative importance of different occupations (including direct and indirect labour), psychological and social conditions of work, and the demands on management.

Integrated data processing and continuous flow operation demand special forms of organisation from design to sales, not just further job subdivision. The possibilities that can be deduced from that fact alone are limited. Pessimists will assume that workers will suffer from additional impersonality and isolation in their tasks. Some even worry that educational requirements will put a large portion of work out of the reach of the "genetic capacity" of most people, thus fostering income inequality and social tension. Optimists, by contrast, hope for a new age of professional commitment instead of bureaucratic and assembly line alienation, for democratic consultation instead of autocracy, and for higher wages and harmony all round.

Oddly enough, some of these vague psychological consequences seem more distinct than the economic results of automation. It is clear that expensive continuous flow equipment yields a higher return when operated around the clock. This means night work, which disrupts family and social life or health, and yields "I don't mind" as the most favourable worker attitude that managers can hope for.[1]

That it is easier to train and manage 50 workers well than 500 poorly is another invulnerable automation epigram. Less obvious is that this inability to handle the many is not so much a fault of the workers as of management's inability to co-ordinate multiple levels of supervision. This inability prevails especially in cultures with strong authoritarian traditions.

The strictly economic consequences have been harder to observe and to predict with accuracy. In the United States one can see that older workers have suffered the most from lay-offs due to automation that makes skills obsolete.[2] But beyond this difference in relative speed of reabsorption

---

[1] F. C. Mann and R. L. Hoffman: "Impact of Automation on Organization and the Individual", in Charles R. Walker (ed.): *Modern Technology and Civilization* (New York, McGraw-Hill, 1962), p. 174.

[2] Richard R. Nelson, Merton J. Peck and Edward D. Kalachek: *Technology, Economic Growth and Public Policy* (Washington, DC, The Brookings Institution, 1967), pp. 123-27.

between the young and the old, there has not been an obvious problem of "structural unemployment", specifically associated with automation, in the United States. Automation has not caused a sharply rising demand for more educated at the expense of less educated workers. More education is available and demanded throughout the American occupational structure. Changes in the occupational composition of the workforce could account for only 15 per cent of higher educational attainment per worker from 1950 to 1960.[1] Moreover, like other new processes, automation raised skill requirements at first but then lowered them in the course of perfection and routinisation. In computer programming, for example, mathematicians with Ph.Ds have been replaced by undergraduates. If automation displaces workers, they may be more educated in one sector, less in another, and either more or less in sectors that produce substitutes or inputs to substitutes for the automated and, therefore, cheaper or better product.

In less developed countries automation has not yet spread widely enough to allow a similar assessment. In some cases automation has little effect on numbers employed. When it was introduced to the Life Insurance Corporation of India numbers were almost unchanged, partly because the company already had a punchcard system. The association of employees had mistakenly feared that 70 to 80 per cent of their jobs would become redundant.[2]

At a higher level of aggregation, Professor Walter P. Krause has weighed the effect of advanced technology in Brazil. He compared trends in productivity, employment, number of establishments, and employment per establishment in technically progressive and lagging sectors, between 1955 and 1959. Labour productivity rose and employment per firm fell in the progressive sectors, but an increase in the number of firms, due to elastic product demand, offset the negative employment effect. In the lagging sectors the number of firms fell but the survivors employed more workers. On the basis of this rather limited evidence, Krause concluded that a widening wage differential, not unemployment, was the real danger of automation in developing countries. Dissatisfaction among less skilled workers in lagging sectors could lead to wage demands that could reinforce inflation.[3]

---

[1] John K. Folger and Charles B. Nam: "Trends in Education in Relation to Occupational Structure", in *Sociology of Education*, Vol. 38, Fall 1964, quoted in Nelson *et al.*, op. cit., p. 143.

[2] The need to relocate employees geographically and reduced future hiring caused the most tension; cf. All-India Insurance Employees' Association: *Automation: LIC's Case X-Rayed* (Calcutta, 1966); International Confederation of Free Trade Unions, Asian Trade Union College: "Automation and Trade Unions—A Case in India Examined in Context", in *Asian Trade Unionist* (New Delhi), Vol. 4, No. 4, December 1966, pp. 21-27; Employers' Federation of India: *Automation—Blessing or Curse?* Monograph No. 10 (Bombay, 1968).

[3] Walter P. Krause: "Economic and Social Aspects of Automation: The Problem of Brazil", in *Revista de Estudos Socioeconomicos* (São Paulo), Vol. 5, 1962, pp. 33-43.

To summarise this subsection, we may say that advanced technology can affect workers both psychologically and more objectively. Whether these effects are, on balance, favourable or unfavourable to a specific type of worker has been difficult to ascertain in developed countries after much experience and study. The evidence for poor countries, including that submitted to this round table, also suggests that the pattern is not simple and obvious.

RESPONSES TO ADVANCED TECHNOLOGY, ESPECIALLY AUTOMATION

High-level management

The remaining part of this paper will examine in sequence the responses of management, labour and government to the characteristics of automation and advanced technology discussed so far. Since automation frustrates one's hopes for a rapid expansion of jobs, even if it does not cause a reduction, and since in some notorious white elephants it may, in fact, have been too complex, large-scale and expensive, the temptation is to criticise any manager who adopts it. One looks for scraps of evidence to show that the villain is grotesquely irrational. Modern economic enterprise, however, hinges on a need to earn profits, and this need lessens the thirst for irrational extravagance in production methods. Since advanced and automated equipment is costly, it is usually not adopted without careful search and critical accounting analysis. There are few occasions when management can compare alternatives as clearly as when equipment is bought, and the decisions are made in haste, emotion, disinterest or without extensive consultation only when someone else is footing the bill. To this extent the investment decision is alike in rich and poor countries, in the twentieth century as it was in the nineteenth and probably earlier centuries.[1] Of course, putting an entire firm on the computer is different from buying automated equipment for one process. Mathematically unsophisticated managers sometimes computerise on faith.

To avoid confusion on this point, however, one must distinguish the practice of management from the enthusiasms of engineers. Max Weber observed that Leonardo da Vinci's "urge was not that of cheapening production but the rational mastery of technical problems as such".[2] Engineers admire elegant solutions, forget cost, and despise tradition. In 1969 a productivity-

[1] That innovations were adopted with the utmost caution in the last century is tested in W. Paul Strassmann: *Risk and Technological Innovation: American Manufacturing Methods in the Nineteenth Century* (Ithaca, NY, Cornell University Press, 1959).

[2] Max Weber: *General Economic History* (New York, Greenberg, 1927), pp. 311-12.

minded Indian wrote that "the chances of a non-computerised society, as against a computerised one, are like those of the professional versus the amateur, in armed conflict. . . . *What this country requires is massive injections of rationalism*, and bona fide computerisation is really the ultimate in rationalism." [1] Investing managers may occasionally sound that way, rejecting the old with undue contempt and viewing the new with undue pride and woolly theories. Perhaps boasting is good public relations; but these managers seldom spend money as extravagantly as words. The following examples show this clearly.

In the Far East (south-east Asia region) only the Japanese and Australian railway systems were employing third-generation computers in 1969. The Indian railway network used second-generation computers primarily for design and development problems and left its yards to manual braking systems, although 1,000 wagons had to be marshalled on some days. The railways in the Khmer Republic, Pakistan, the Philippines and Thailand used no computers at all.[2]

When the Mexicans built an automated sponge iron plant in 1960, it was hailed as the very latest thing, and all asserted that the country need accept nothing less. Later, Mexico found it wise to install three open hearths when the rest of the world has shifted to oxygen converters. The focus of publicity shifted from technology to capacity. The sponge iron plant had been made a showpiece of automation to enhance its appeal to visitors. The process had been invented in Mexico and the company wished to license it throughout the world. Automation was partly a symbol of modernity.

As a matter of fact, there is today an undue horror of investment in symbolism. The implication is that any attention to the visual, symbolic effect of an installation and its working must detract from its efficiency. These assertions are usually made without any reference to what is being symbolised and communicated. The apparent assumption is that symbols must represent power and prestige. But modern installations lend themselves better to symbolising teamwork efficiency and avoidance of waste. If the symbol conveys clearly that everything unnecessary—the ornamental, the sentimental and the traditional—has been cut away with relentless economy, then one undermines those "anti-productive attitudes which poison the whole social organism, forming a barrier to the improvement of work organisation and methods, as well as to automation".[3]

---

[1] D. H. Bhutani: "Computerisation II", in *Productivity* (New Delhi), Summer 1969, pp. 1 and 4. Italics in original.

[2] United Nations, ECAFE, op. cit., p. 64.

[3] Gabriel Ardant, "Automation in Developing Countries", in *International Labour Review*, Vol. XC, No. 5, Nov. 1964, p. 433.

One of the most careful tests of the technological rationality of management was made for textiles by the Economic Commission for Latin America (ECLA). Vintage equipment for 1950, 1960 and 1965 was studied for 11 processing stages: opening, carding, combing, drawing, slubbing, roving, spinning, winding, warping, sizing and weaving. Although the annual costs for the 1965 equipment were 60 per cent above that for 1950 and 16 per cent above that for 1960, total unit costs of production were 9 per cent and 3 per cent less.[1] The Commission did not estimate what costs might have been if the older equipment had been purchased second-hand. These calculations have been made since then by Professor Dilmus James, who has found that the 1950 vintage would remain uncompetitive but that the 1960 vintage would be competitive if basic equipment cost fell by 50 per cent; 25 per cent would be insufficient.[2]

Where managers waste resources through premature automation, the "irrationality" is most probably due to the signals conveyed to them by the market as influenced by government action. A striking example comes from Brazilian brewing. In Rio and São Paulo, two important breweries are still operating with the technology of the 1930s. During the early 1960s the two companies were induced to set up branches in the lagging north-east where wages are much lower than in Rio or São Paulo. But the two new plants were highly automated, especially in materials handling: the Government's investment incentive scheme had subsidised plant and ignored important employment effects.[3]

A final comment on management rationality needs to be made. There may be cases where labour-intensive methods seem to be rejected prematurely. But if the switch has to be made eventually, and if length of experience bears heavily on efficiency, then an earlier switch may possibly pay off if the experience with the labour-intensive methods is certain to prove irrelevant. In this case, the divergence of practice from the simple model of factor-cost and scale

---

[1] United Nations, ECLA: *Choice of Technologies in the Latin American Textile Industry* (New York, doc. E/CN.12/746, 13 Jan. 1966; mimeographed). This report, prepared for the Latin American Symposium of Industrial Development (Santiago, Chile, 14-25 Mar. 1966), does not consider fully automated techniques which have become available. Moreover, 3 possibilities for 11 steps would yield a total of 177,147 possible combinations if equipment from different vintages were to be mixed. Some of these might be cheaper than the straight use of 1965 technology.

[2] Dilmus James: *Used Machinery and Less Developed Countries*, unpublished doctoral dissertation, Michigan State University, 1970.

[3] Corporations that invested in approved projects in the north-east received 50 per cent income tax reductions (the "34/18" funds). New projects have a capital/labour ratio almost four times the average of that prevailing in 1959, and industrial employment actually fell by 14 per cent from 1959 to 1965. See Otto G. Wadsted: "A Industrialização do Nordeste: Alguns Aspectos de Longo Prazo", in *A Economia Brasileira e Suas Perspectivas*, July 1968, pp. 241-59.

means that management, taking a longer view, is more rational than the over-simplified model.[1]

Labour

The response of management to automation has been discussed in the preceding subsection on the basis of two criteria: raise sales and avoid waste. The assumption is that, with correct market signals, higher profits mean faster development and less unemployment because technology of the right capital intensity will have been chosen. The response of labour cannot be judged so simply.

First, we cannot say that the response of labour must be judged by its effect on development, assuming that growth of the gross domestic product is a synonym for the self-interest of today's labour force. Higher wages now, at the expense of growth, are a valid choice.

Second, the net self-interest of the whole labour force—that is, everyone working or seeking work—is different from that of employees in any particular plant, taken in isolation. The trade union spokesmen for these employees will, in turn, have goals that only partially overlap those of union members. By contrast, top management is expected to represent neither industry in general nor junior executives. Only the profits of the firm count.

Third, the interests of labour are not purely monetary. Participation in production is a way of life with human considerations that go beyond sales and the payroll. The bitter opposition to Frederick Taylor's scientific management was not only economic self-interest but also opposition to "the concept of the worker, not as a human being, but as a mere instrument of production, to be used with no more consideration of his humanity than any other tool".[2] While the initial and ultimate psychological effects may be identified and predicted with surprising accuracy, their relative importance cannot easily be gauged in economic terms.

Finally, one must choose whether to judge the response of labour by statements or action. Which distorts the true feelings of labour the least? Speeches and articles by labour leaders and their staffs are rarely objective appraisals in a quest for truth but rather part of the bargaining game. The contract that is finally accepted reflects what could be attained, not necessarily what was sought as best. For management the criterion is clear: it seeks the lowest wage that attracts and retains an adequate labour force—the lower

[1] See the concept of "localised technical progress" in A. B. Atkinson and J. E. Stiglitz: "A New View of Technological Change", in *The Economic Journal* (London), Vol. LXXIX, No. 315, Sep. 1969, pp. 573-78.

[2] Gerald G. Somers, Edward L. Cushman and Nat Weinberg (eds.): *Adjusting to Technological Change* (New York, Harper & Row, 1963), p. 208.

the better. Moreover, the response of management is not judged by chamber-of-industry rhetoric, nor by labour-contract compromises. If management takes steps to introduce automated or computerised methods, that and nothing else counts as a sign of preference.

Historically, the response of labour to technological change has usually been indifferent or negative. An exception arose in the textile and machinery industry of New England in the early nineteenth century. Here workers took to inventing for their employers with such enthusiasm that by mid-century the British marvelled at "Yankee ingenuity" and sent Royal Commissions across the Atlantic to seek technical assistance. Meanwhile, British workers could not understand why the Americans were not afraid of inventing themselves out of jobs.[1] I can think of no other time or place (though they may exist) where the workers went beyond docile co-operation with management and on a large scale took the initiative in promoting better methods.

Where an innovation saves labour compared with an existing activity (which is not always the case), the more common reactions on the part of workers can be classified as follows:

— violence to the inventor or innovator (reaction 1);

— destruction of the novel equipment (reaction 2);

— preventing use of the novelty (reaction 3);

— permanent slow operation and over-staffing (feather-bedding) (reaction 4);

— opposition to lay-offs, often meaning transitional over-staffing (reaction 5);

— demand for compulsory consultation on reorganisation (reaction 6);

— demand for increased compensation for employees and assistance for those laid off (reaction 7).

In Western industrialised countries the trend has gone from the more destructive reactions at the top of the list to the more tempered ones further down. The trend reflects not only experience with innovations and with effectiveness of alternate strategies by labour, but changes in style throughout the culture. For example, when judging the strangling of the inventor of a weaving machine in Danzig in 1579, one should recall that at that time capital punishment was a popular method of hindering religious heresies, witchcraft and novel ideas in general.

The tactics, common around 1800, of destroying equipment instead of inventors show that the right to life of individuals and ideas *per se* was no

[1] For the Report of the Committee on the Machinery of America, 1855, and copies of other historical documents, see Nathan Rosenberg (ed.): *Great Britain and the American System of Manufacture: A Study in the International Transmission of Technology* (Chicago, Aldine, 1969).

longer questioned. •At the same time, methods for the exchange of ideas among individuals of different economic classes were rather imperfectly developed. In the course of the nineteenth century, refinement of property rights tended to prevent physical damage to equipment as factory owners gained political power. Workers had to persuade owners and government that they were capable of bargaining peacefully about whether, how and at what rates of pay new equipment would be used. Violence was assigned to the political sphere, showing that, with the division of labour between politics and economics, the culture had reached a higher level of differentiation. It was inevitable, perhaps, that in this cultural setting workers would emulate owners by cradling a sense of property rights not only in their jobs but even in work rules. Where these rules were deeply embedded and elaborate, as among American railway workers and longshoremen, technological change still depends on evolving ingenious processes of accommodation, meaning one form or another of compulsory consultation on reorganisation.[1]

In general, however, this proprietary fixation by some workers is considered atavistic in the more sophisticated unions, such as the automobile and rubber workers in the United States. In a technocratic age, they do not question the organisational talent of management. They bargain for increased compensation in various forms for those who stay and for help with retraining, placement and mobility for those who go. In these matters Sweden appears to be most advanced.

One should note that the seven reactions listed above have been discussed in the sense of aims perceived by spokesmen for labour. In the process of attaining a more advanced aim, the unions may nevertheless use tactics that resemble earlier aims. To gain higher compensation, workers may slow down or obstruct the use of a machine. A bargaining process may break down and lead to violence. These transitional phenomena should not be confused with labour's primary objectives.

In less developed countries the response of labour to technological change, and especially automation, will not follow in sequence the responses made in developed countries. On the one hand, the cultural and political setting will be different and on the other, developing countries can draw on the labour-relations experience of advanced countries. There may be violence among vested interests and challengers, but this violence will not be covered by reaction 1 or 2, which are aimed against new technology and its sponsors *per se*.

To bargain for the points that are bargained for in advanced countries

[1] Charles C. Killingsworth: "Co-operative Approaches to Problems of Technological Change", in Somers *et al.*, op. cit., pp. 61-94.

(reactions 6 and 7) will seem alert and up to date, but some adjustments to local needs are unavoidable. Compared with the situation in the West in the nineteenth century, management will usually be weaker and government stronger but less effective, and they may represent different social and ethnic groups from labour and from each other. These differences in knowledge and power help to account for the differences in goals.

The most common goal, as most case studies show, is reaction 5, i.e. insistence on temporary over-staffing. The standard of living of no worker currently employed by a company is to decline. A firm may reduce employment to the level needed after automation only by waiting until enough workers have voluntarily retired. At times, it may close a plant altogether, however, and open a new one in a different town, perhaps after a change in name. Here the attitude of government is crucial. There are cases, such as occurred in Cuba in the 1950s, where reaction 4, "permanent slow operation and over-staffing (feather-bedding)" were labour policy. More typical is the guarantee obtained by the Indian Railway Union that staff made redundant by computers would not be discharged.[1]

Reaction 3, "preventing use of the novelty", remains to be considered. Although not the most typical, this reaction nevertheless occurs in certain trade union settings. Oddly enough, the most likely settings are either when a union is strong and militant up to the national level or when unions completely fail to dominate an industry and must compete with one another for membership. In the first case, the single strong union may force support at the plant level and make a public appeal to broad ideologies and to general employment opportunities. A setback for a corporate management, public or private, that supports an opposition political party is a gain for this kind of union and its political allies.

In the other case, competition is not only along class or political lines but also among the unions themselves. Where several unions compete for membership within a plant, the most intransigent may not necessarily have the most appeal, because each worker will know how a proposed installation change is to affect him. But if the national association must compete with others, the most intransigent line does lend itself best to its membership campaigns outside the plant.

A combination of the two cases can be found in the Mexican electric power industry. This will be summarised in the next subsection on the response of government. Here the attempt to prevent use involved not only competing unions in a newly consolidated industry but also the conditioned reflex of a patriotic confrontation with a foreign management.

---

[1] United Nations, ECAFE, op. cit., p. 28.

In most developing countries, however, the existence of a large number of unemployed, who are eager to replace anyone now on the production line, weakens union strength, which therefore depends on laws and government support. The government response is reviewed in the next subsection, but preliminary comments may be made here.

Sophisticated economists advise governments to make policy statements insisting that spending on automation, even if favoured by everyone at the plant level, must consider the social opportunity cost of the resources. In practice, however, machinery is not set up to define what this means and to enforce the adoption of social over private goals. Usually government will try to avoid conflict situations and be satisfied if management and labour have consulted and reached an agreement about retraining some workers and compensating others. Reducing conflict is perfectly consistent with lowering social opportunity cost if that term is defined with sufficient breadth. Unfortunately, that step goes beyond what is measurable. Thus we find that one government response is to nudge labour toward reactions 6 and 7, consultation and compensation. The Mexican Government has such a policy. According to the current labour law, it appears that any plant modernisation is allowed if workers accept it with a formal agreement. Only in the absence of such an agreement will the Board of Conciliation and Arbitration step in. In either case, retrenched workers are entitled to a minimum of 4 months' pay plus 20 days' pay for each year of service. But since government is concerned with more than worker reaction and welfare, we should now examine its response from all angles.

Government

The responses of government to automation in developing countries, like those of labour, can theoretically range from violent opposition to grudging or enthusiastic acceptance. Prevention of use, destruction, or violence to the innovator are not options that are widely used. At the other end of the scale, however, outright enthusiasm is also rare.

Most governments understand that social processes, including wage bargaining, function more smoothly if total product is expanding. In this sense, governments favour and support technological change. But when it comes to expensive computers and automated equipment, doubts arise. Will there be a net gain in output proportionate to the turbulence of dislocation?

How governments view these costs and gains depends on who is in power. A governing group representing large landowners will view the net advantages of a more or less drastic reorganisation of agriculture differently from one representing landless workers, or mainly peasants, or mainly urban dwellers of different ethnic origins. Similarly, constraints on automation will vary if a government represents small-scale competitors, labour, the military, or large-scale industrialists.

But self-interest, as expressed through the political process, is not the only important variable. The response depends also on the ability of a group to *perceive* how automation is related to its self-interest. Such perception is clearly harder for government than for workers who are about to be displaced or retrained and advanced.

To simplify the discussion, we may say that automation may have both beneficial and harmful consequences. Government may foresee these accurately, or not at all, or it may underestimate or overestimate the results. Since there are 4 basic possibilities for the benefits, and 4 for the harms, a theoretical maximum of 16 combinations exist. But only 6 seem sufficiently widespread and interesting to call for close examination. These are as follows:

— ignorance: the government is not aware that anything noteworthy is happening and foresees no significant public gains or losses;

— marksmanship: accurate prediction of consequences;

— pessimism: gains underestimated, harm overestimated;

— optimism: harm underestimated, gain overestimated;

— phlegmatism: both underestimated;

— perplexity: both overestimated.

Historically, the reaction to innovations has followed an ignorance-phlegmatism-pessimism-optimism sequence. Where innovations are gradually contrived, their appearance may not be startling. Until the late Middle Ages this gradual, obscure evolution of new production methods was the most common case. Exceptions were innovations introduced from distant lands by armies or travellers. Even in modern times, some innovations have been too intangible and invisible to penetrate the understanding of government officials. Such was the case after 1800 with the system of precision metalworking that allowed the mass assembly of machinery with interchangeable parts.

Phlegmatism has been most common in advanced countries until the middle of this century. From the cotton gin to the automobile, a few costs and gains were foreseen, but insight fell far short of actual ramifications through the economy and society. The same was true of an early example of technical assistance—the introduction of the wheel to the Papago Indians in 1900. The Indian agent of the United States Government thought that transportation would be a little more efficient among the desert villages in his care. Instead, a major occupational and social transformation occurred.[1] Phlegmatism was least likely where harm and benefit had military value.

Pessimism about technological change is usually the perception of authorities who view their own position in a society as beyond improvement. Novelties that create uncertainty and possibly conflict among lower groups could only unsettle a presently satisfactory state. Thus the mediaeval Church opposed printing and the clock, among many other novelties, and thus the Tokugawa sealed Japan against Western influences. Outright pessimism about technological change is no longer a position that is openly held in underdeveloped countries. In advanced countries people worry about pollution and the exhaustion of fossil and other fuels, and about cultural effects, to the point of net pessimism about technology. In lands of intense squalor and hunger such views are not plausible.

A more likely reaction in poor countries nowadays is one of perplexity. The penalties seem great where installations are novel, expensive, but in need of few workers. Yet the gains in output and efficiency cannot be sacrificed. Governments feel damned if they allow it, and damned if they do not. Their position papers take on a tone of perplexity. This tone may, however, be largely borrowed from foreign development experts who seem even more worried about the employment problems than the governments they advise. Perhaps this more intense reaction is due to the foreigners' inability to distinguish élite and non-élite persons in a strange culture and to discount unemployment among the non-élite. The first perplexed view of technological change held by a principal leader of a recently freed colonial territory was that of Thomas Jefferson. He had always viewed factories as a corrupting force in society, but during political difficulties with Britain in the early 1800s, he began to fear that the delayed introduction of factories to the United States was also a menace.

With respect to automation, in particular, perplexity seems to have been lowest in those developing countries that have shown least respect for Western opinions and policies: the Arab countries of the Middle East. At a 1967 conference on scientific and technological prospects for this area, automation was mentioned only once. An oil consultant mentioned that per million tons of oil refined, the latest European refinery used only 20 instead of 2,000 to 4,500 workers (as in the Middle East). No one seemed alarmed by the trend, and no discussion ensued.[2]

This phlegmatic reaction should be contrasted with the conclusions of a 1968 New Delhi UNESCO Conference. Industries were encouraged

[1] W. L. Bliss: "In the Wake of the Wheel", in Walker, op. cit., pp. 274-80.

[2] Ghanim Alukaili, in Claire Nader and A. B. Zahlan (eds.): *Science and Technology in Developing Countries* (Cambridge University Press, 1969), p. 296.

to develop their own more labour-intensive technology, and to be given preferential treatment, *but only* in cases where they reached a "comparable nature, efficiency and level [to] the alternate foreign technology".[1] The Indians were doing their best to learn from experience, but what did experience teach? Of three famous steel mills built in the 1950s at Rourkela (German), Durgapur (British) and Bhilai (Russian), the most modern and automated was the one at Rourkela. At first it seemed the least efficient to the point of being labelled "sick", but by the mid-1960s it was the most profitable and productive, exceeding rated capacity by 7 per cent.

The Colombian Government, among others, has expressed a similar wary attitude towards automation. One spokesman of the National Planning Department said that the Government would systematically promote the use of labour-intensive methods and would intervene to moderate the degree of automation. Machinery and equipment would be limited to "those projects where technical conditions and urgency make such use necessary". The spokesman realised that "modified automation" for more jobs required wage restraint on the part of labour.[2]

But as soon as the process of development acquires momentum, usually with a favourable balance of payments and with government in the hands of pragmatic technocrats, the attitude towards automation changes from perplexity to optimism. One might have thought that the experience with costly textile mills, inefficient automobile assembly and white elephant steel mills would have had a sobering effect on economic planners. But these disappointments, quite properly, have not turned them against the latest technology as such, but against its small-scale, internationally uncompetitive application. To be in shape for export competition, the steel plant set up in Ciudad Guayana (Venezuela), for example, had to be re-equipped with the latest technology.[3]

An interesting case that shows a government promoting automation with great determination is that of the Compañía de Luz y Fuerza del Centro, the foreign-owned light and power company serving Mexico City. Clause 30-III of the company's contract with the union, the Sindicato Mexicano de Electricistas (SME), specified that no change in the quantity or quality of any worker's job could be made without renegotiating his wage. Before 1960 the union not only refused to allow installation of a computer for billing but also

---

[1] UNESCO, Conference on the Application of Science and Technology to the Development of Asia, New Delhi, 9-20 August 1968: *Final Report*, Part I: *Conclusions and Recommendations* (Paris, 1969), p. 20.

[2] The Economist Intelligence Unit Ltd.: *Quarterly Economic Review: Colombia, Ecuador* (London), 1969, No. 1, p. 6.

[3] Lloyd Rodwin (ed.): *Planning Urban Growth and Regional Development: The Experience of the Guayana Program of Venezuela* (Cambridge, Mass., MIT Press, 1969), pp. 168-69.

objected even to reducing meter readings and billings to a bimonthly basis. It insisted that employment should grow in proportion to the number of customers, but that no new "employees of confidence" could be hired. Wages and fringe benefits rose faster than in the rest of the electric power industry and, together with a lagging rate structure, led to uneconomic operations. This difficulty was disguised through power purchases at subsidised rates from the Federal Electricity Commission. The International Bank was reluctant to finance the expansion of the entire system on this basis, and so a substantial part of the Mexican development plan seemed threatened.

In 1960 the Government nationalised the company, together with other parts of the power system that it did not already own. Raising rates, integrating the system on a national single cycle and political symbolism all contributed to this decision. Higher efficiency and the introduction of automation, however, were also among the Government's objectives, and were carried out when the President himself forced acceptance after a challenging series of slow-downs by the union. The union, on its part, received a number of concessions, including provisional retention of clause 30-III and a guarantee of no demotions or pay reductions. But for the sake of efficiency, the President insisted that the experience of nationalised petroleum, with worker participation in management, would not be repeated.[1]

CONCLUSION

The response of labour to automation in developing countries has centred on its immediate labour-displacing effects. The response of management has centred on the higher profits that can be earned from automation. These profits are higher not only when the equipment and its operation add less to costs than the displacing of workers subtracts. The automated or computerised process may save capital as well as labour. It may allow increases in output without spending more for installations like railways, canals, power lines and warehouses that can be used to greater capacity. Automation may raise profits and sales through better quality and greater variety in products or services. The real importance of these factors compared with pure factor-

---

[1] See Mark Thompson: "Collective Bargaining in the Mexican Electric Industry", in *British Journal of Industrial Relations* (London), Vol. VIII, No. 1, Mar. 1970, pp. 55-68; idem: "The Development of Unionism among Mexican Electrical Workers", unpublished doctoral dissertation, Cornell University, 1966, pp. 222-24, 290-96. I am grateful to Dr. Thompson for additional observations made in the course of several conversations. See also Miguel S. Wionczek's analysis of the industry in Raymond Vernon (ed.): *Public Policy and Private Enterprise in Mexico* (Cambridge, Mass., Harvard University Press, 1964), pp. 19-110.

substitution stands out only if labour, capital and foreign exchange are correctly priced.

The response of government to automation is the least predictable since, unlike labour and management, it has conflicting goals. Government wants to avoid short-run political problems by not depriving management of sales and profits and by not depriving workers of jobs. It wants to avoid intermediate political problems by fostering economic growth, efficiency and investment without inflation and balance-of-payments crises. These are conflicting goals that cannot be compromised without compromising the aims of specific economic interest groups. The relative political strength of these groups, including the extent of their representation in government, determine in part how government perceives the merit of encouraging or delaying automation.

The responses of governments in developing countries to automation are different now from the responses to better technology in industrialised countries in the last century. The difference is probably not due to any unique technical features of automation. In their own way, nineteenth-century innovations also fitted larger volumes, raised quality, heightened variety and displaced labour (although quality and variety losses also occurred). The main difference between then and now lies in political conditions. The workings of economic systems are now understood somewhat better, and governments view themselves as responsible for seeking growth without crises. Governments are unlikely to be ignorant or phlegmatic about technological change. At the same time, they must tolerate more pressure from organised labour than in the nineteenth century. This pressure is likely to be greatest against measures that promote efficiency and growth when growth is least. In recent decades automated equipment and computers have become a symbol of this conflict between efficiency and employment. It is therefore not surprising that the attitudes of some governments towards such equipment could be described as nervous. This attitude can be expected to change to optimism as economic growth takes on momentum. Both attitudes are better than the pessimism of former vested interests towards innovations that led to smashed machines and murdered inventors.

TECHNOLOGY, EMPLOYMENT AND GROWTH: LESSONS FROM THE EXPERIENCE OF JAPAN

GUSTAV RANIS
Yale University

Granted that the over-all growth performance of the less developed countries in the 1960s was substantially better than that in the 1950s, there can be little doubt that the biggest crisis lies just ahead. This is so partly because (as more and more people are beginning to recognise) progress has been very unevenly distributed, and partly because the threat is of an acceleration of this trend in the 1970s and 1980s. Perhaps the most important manifestation of this uneven participation in the past has been that, even in the fastest-growing countries, unemployment and underemployment rates have been rising. Second, all available "guesstimates" and projections for the future seem to agree that even if population growth could be substantially reduced tomorrow, a labour force explosion of major proportions must be expected in the less developed countries over the next decade or so in view of the age structure of the present population. Add to this the fact that the volume of foreign aid and of foreign private capital both available and acceptable in the 1970s is, in spite of all hopes, pleas and efforts to the contrary, likely to fall substantially below that of the 1960s, and the true dimensions of the problem ahead become clear. If major political as well as economic crises are to be avoided, the less developed countries are going to have to solve their future output problem somehow, yet not at the expense of employment and distribution; and this will have to be accomplished largely by their own efforts.

During the 1950s and early 1960s most of the less developed countries engaged in what have been called "import substitution policies". These usually included, in one package, a by now well-known syndrome of policies: exchange controls and import licensing, budget deficits, overvalued exchange rates and low (sometimes negative) real interest rates. The aim, generally speaking, was to redirect pre-independence traditional colonial flows towards social and economic overheads and import-replacing consumer goods industries. The consequences of this set of policies on economic performance have by now been fairly well recognised and acknowledged: a spurt of inefficient industrial growth (capital- and import-intensive) accompanied by a discouragement of exports and agricultural output, low domestic saving rates, a relatively heavy dependence on foreign aid, and low rates of technological change.

As governments in the less developed countries became increasingly aware of the economic cost of these policies, one could observe, during the 1960s, a growing tendency to move towards a new policy package. This package can be described, at the cost of some oversimplification, as the readjustment of a number of crucial, previously distorted, relative prices, including the exchange rate, the interest rate and the internal terms of trade. The governments replaced quantitative controls in the foreign exchange market with tariffs, and moved towards more realistic exchange rates, via either a *de jure* or *de facto* devaluation. With a relatively free market and higher interest rates, they replaced the severe credit rationing and the forced procurement of food at artificially low prices. Participation in development could therefore be offered for the first time to medium- and small-scale entrepreneurs in both agriculture and industry. The effects of this type of restructuring, where it has been carried out at least in part, as for example in the Republic of Korea and in Pakistan, have indeed been remarkable in turning situations of virtual stagnation in the 1950s into sustained growth situations in the 1960s.

More specifically, once agriculture is no longer discriminated against by unfavourable terms of trade, this sector can play its historical role of generating surpluses which, when successfully channelled, can provide employment opportunities for the unskilled labour being released, a more broadly based industrial development pattern using relatively more domestic materials and a more labour-intensive technology, and exports (especially of the non-traditional, labour-using variety). Domestic saving rates can move up into the Rostow take-off range, and indigenous technological change can assume much greater importance.[1]

Perhaps the most important fact from our point of view here is that the new signals induce the adoption of different, more labour-using and unemployment-reducing technologies and output mixes. In this context the vital role, for better or worse, of technological flows between rich and poor countries must be kept in mind. The very coexistence of countries at widely different levels of technology undoubtedly represents one of the most important past, present and prospective influences on the performance of

[1] For a fuller discussion of the typical import substitution phase of development in the less developed countries and of the transition to a more efficiency-oriented phase, see Gustav Ranis: *Relative Prices in Planning for Economic Development* (New York, National Bureau of Economic Research) (in preparation).

the less developed countries. It is the precise nature of these technological flows and the way in which they have been accommodated by the countries in question that has, in my view, had a decisive impact during these past two decades of development. To put it another way, it is also in this area that the greatest potential for improved performance by the less developed countries in the 1970s can and must be located.

The move from an import substitution to an export substitution dominated growth pattern that began in the middle 1960s, and the consequent marked changes in economic performance, are still the exception and not the rule in the less developed world. In spite of the demonstrations of what can in fact be accomplished, there remain formidable obstacles to the dismantling of the import substitution régime. Direct controls imply absolute power (as well as supplementary incomes) for the civil service, which it is loath to surrender. Moreover, the inevitably greater role for private enterprise under any liberalised régime runs up against associations with colonialism and fears of anti-social concessions.

In addition to this pull of vested interests and some quite well-intentioned doubts concerning the risks of liberalisation, there remains a good deal of scepticism concerning the major role we have accorded here to technological change as a determinant of success in development. In particular, many officials in less developed countries, many aid donors and many scholars share the view that most technological change (especially outside agriculture) must take place abroad, and that the borrowing countries have in fact only a very narrow set of technological choices open to them. If only the latest vintage machinery of advanced countries is relevant, all the talk about alternative factor proportions in response to alternative resource endowments necessarily becomes irrelevant—or restricted to changes in output mixes via trade.

Scepticism on both of these points—the merits of abandoning import substitution, and the scope of technological choice—is of course not unrelated; for if there is no real alternative to large-scale capital-intensity, perhaps the most powerful argument for changing the basic policy package loses much of its force. The rest of this paper will therefore concentrate on suggesting, in the next section, a more realistic view of the nature of the innovation process in the borrowing developing countries. The empirical relevancy of this view is then explored in the last section.

EXPLOITING THE TECHNOLOGICAL SHELF

There is less doubt now than ever before that the success of a development effort is likely to be related much more to technological change than to the growth of physical inputs. Nevertheless, in spite of this acknowledged importance of technological change, it has been difficult to achieve a clear understanding of the process by which innovations are actually made in a typical developing or borrowing country.

First and foremost, it must be remembered that the nature of technological change is different in the two parts of the world. In an advanced country, technological change is viewed as in the main automatic and routinised, or as capable of being generated through "research and development" expenditures according to some rules of cost/benefit analysis. In contrast, we know that in contemporary developing societies, technological change can be neither taken for granted nor afforded through "R and D" allocations. In this situation we cannot avoid the question of what, given the existence of a shelf of technology from abroad, is the pattern according to which the typical less developed economy, in fact, manages to innovate. This question in turn forces us to look at least at the following dimensions more carefully: (*a*) the precise nature of that technology shelf; (*b*) the availability within the less developed countries of required initial managerial and entrepreneurial capacity; and (*c*) the changing nature of that required managerial and entrepreneurial capacity in the course of transition to modern growth.

The technology shelf developed in mature industrial economies may be described by a set of unit activities following a smooth envelope curve (figure 1). A particular technology can be described by an L-shaped contour producing one unit of output with a given pair of capital and labour coefficients. The technology shelf is composed of the complete set of such activities or technologies which have been demonstrated to be feasible somewhere in the advanced countries at some historical point in time, including the present. Since there exist a number of technology-exporting countries (for instance, Federal Republic of Germany, Japan, United Kingdom, United States) with continuous technological transfers amongst themselves as well as with the less developed countries, it is not unreasonable to postulate the existence of a single technological shelf for the lending world as a whole. For example, unit technology A_0 may have been generated in Germany in 1920, A_1 in the United States in 1920, A_2 in the United States in 1950, and so one. In other words, as we move to the left along the shelf, we run into more modern technology, i.e. technology of more recent vintage and of higher capital-intensity. As capital per head increases this means that the typical worker has learned to co-operate with more units of capital of increasing technical complexity. This capital-deepening process, in other words, is more complicated than the textbook version of "homogeneous" labour being equipped with more units of "homogeneous" capital.

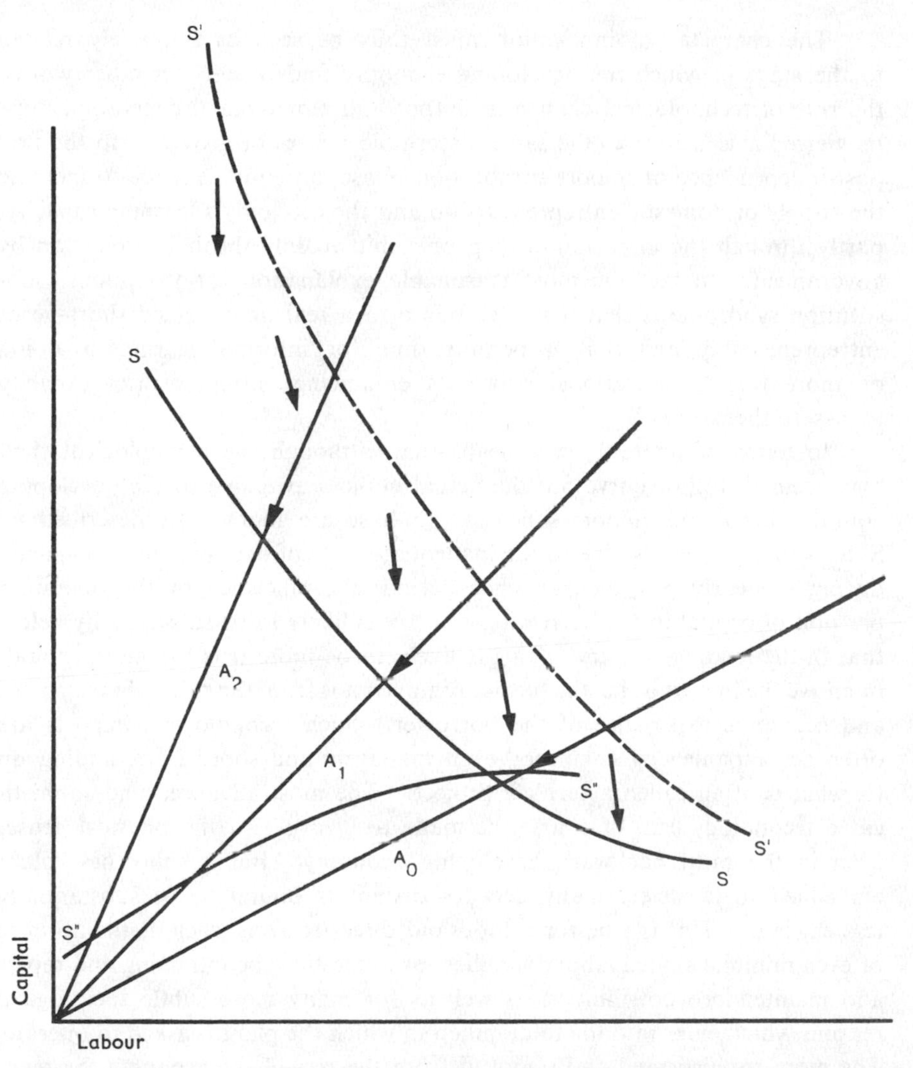

Figure 1. Output per unit of capital and labour: the "technological" shelf (for explanation of symbols, see text)

At any point in time the typical less developed country is then theoretically free to borrow a particular unit activity from anywhere along this shelf. What technology is chosen and what happens as an immediate and ultimate consequence of that choice, that is to say, what secondary processes and reactions are set off, is of course all part and parcel of the innovational process taken as a whole. The quality of each step of that process in turn depends on the nature of the entrepreneurial, managerial and skilled labour capacity of the borrower.

The character of innovation must thus be seen as intimately related to the stage in which the developing economy finds itself. In other words, the role of technological change in output and employment generation must be viewed as sensitive to the same discernible phases of growth. In the first post-independence or import substitution phase, an effort is made to increase the supply of domestic entrepreneurship and the economy's learning capacity, partly through the importation of people, but mainly through protection by government. In fact the most reasonable explanation for the import substitution syndrome is that it is a response to a real or imagined shortage of entrepreneurship and that it permits time for informal learning-by-doing or more formal educational processes enhancing entrepreneurial capacity to assert themselves.[1]

In terms of figure 1, this means that, although the technological shelf may seem to follow curve SS, the actual choices available to the developing country during the import substitution phase are more aptly described by S′ S′. In other words, due to the inadequate state of entrepreneurial capacity during the early post-independence period, the efficiency of the operation per unit of capital in the borrowing country is likely to be substantially below that in the lending country. This is likely to be more true the more capital-intensive the import is, i.e. the further removed it is from the cultural inheritance and economic experience of the borrower. Such technological imports are often accompanied by engineers, even managers and supervisors, adding up to what is often called a turnkey project. The most advanced and sophisticated technology can, of course, be made to "work", in the physical sense, even in the most backward developing economy. But a shiny new plant embedded in a society many decades distant is bound to be substantially less efficient. This is true for a thousand direct reasons, such as the absence of even minimal skilled labour supplies, of domestic subcontracting and repair and maintenance possibilities, as well as for many more subtle sociological reasons which enter into the total milieu in which the plant is asked to operate. The more sophisticated and removed from the rest of the economy the technological transplant is, the greater will be the relative inefficiency, as indicated by the shape of the S′ S′ curve.

If and when the economy moves away from the import substitution phase and enters the second phase of liberalisation and export promotion, a second important, if unintentional, type of innovation is likely to make its

[1] A few countries, such as Malaysia, with command over a very strong and reliable natural resources base, may be able to avoid such a phase altogether. Moreover, there clearly exist better and worse (i.e. less and more costly) import substitution packages to choose from (for example, compare Brazil and Ghana); we cannot, however, expand on this very interesting subject in the context of the present paper.

appearance, namely a reduction in the extent of the inefficiency of the original transplanted technology. Call it X-efficiency if you like, but the cost of the pure transplantation is likely to be reduced, quite unintentionally, largely as a result of factors external to the profit-maximising behaviour of the productive unit itself. This increase in productive efficiency will increase in quantitative significance as the import substitution hothouse temperature is gradually turned down and a more competitive economy emerges. In figure 1 the effects of gradual enhancement of efficiency may be represented by the arrows tending, in time, to move S′ S′ back towards the original SS position.[1]

Another more conscious and quantitatively more important type of innovation begins to gather importance during this same second phase of transition. This phenomenon may be called innovational assimilation, that is, innovating "on top of" imported technology in the direction of using relatively more of the abundant unskilled labour supply. As the economy shifts from a natural resource-based growth pattern in the import substitution phase to a human resource-based pattern in the export substitution phase, this means an increasing sensitivity to the continuously changing factor endowment, first in terms of the efficient utilisation of the domestic unskilled labour force and later in terms of the incorporation of growing domestic skills and ingenuity. In other words, the appropriate type of technology finally in place must be one in which not only the initial choice from the shelf, but also the adaptations and adjustments consciously made thereafter in response to changing domestic resource and capability constraints, play an important role.

The more liberalised the economy (in terms of the government's performing a catalytic role through the market by indirect means, rather than trying to impose resource allocation by direct controls), the better the chances that the thousands of dispersed decision makers can be induced, by the sheer force of profit maximisation, to make the "right" decisions. Even in the absence of technological change, as long as surplus labour overhangs the market (and the expectation is for even more of this in the future) we can expect only modest—exogenously caused—upward movement in real wages and in the capital/labour ratio. Superimposed is the above-mentioned assimilating type of innovational behaviour which tends for the same reason to be slanted in the labour-using direction. In the typical labour surplus type of economy, or in one likely to become so over the next decade (as is probably the case in much of Africa), all this means as much efficient accommodation of pure

[1] A more sophisticated analysis, differentiating between initial labour and capital inefficiency and between the labour- and capital-saving nature of this late type of innovation, is possible but will not be introduced here.

labour services as possible.[1] Whether this will lead to a sectoral output shift in favour of labour-intensive export commodities or a mix predominantly addressed to the domestic market depends, of course, on the size and other characteristics of the economy. No strong generalisation as to the relative importance of shifts in output mix as against changes in technology for given mixes is likely to be valid. It should be clear, however, that the important issue is that the search for innovation can now be considered a conscious activity of the individual entrepreneur and—given the combination of more realistic relative price signals after liberalisation and greater entrepreneurial capacity—that it is likely to be directed towards various forms of indigenous capital-stretching on top of the imported technology. Such capital-stretching can be represented by a reduction in the capital coefficient per unit of output. The effective post-assimilation set of unit technologies (i.e. after domestic assimilation) may thus be represented by curve S″ S″, with the strength of the indigenous labour-using innovative effort indicated by the amount of the "downward" shift in the capital coefficient.

It should be noted here that a negatively sloped technology shelf (e.g. SS), representing pure technological transplantation, permits, as the economy moves to the left, higher labour productivity levels, but only at increasing capital cost. In a country characterised by capital scarcity this may mean increased technical unemployment (*à la* Eckaus) and hence a lower value of income per head for the economy—in spite of the higher level of labour productivity achieved. Domestic capital-stretching, however, can materially affect that situation by enabling more workers to be employed per unit of the capital stock. If the post-assimilation unit technology set (e.g. S″ S″) is upward-sloping, as the economy moves to the left by first borrowing abroad and then innovating domestically on top of that borrowed technology, higher labour productivity levels become consistent with lower capital/output ratios.

We may summarise the situation by stating that once the over-all policy setting, as described in the previous section, has turned more favourable and permitted the economy to enter the second phase of transition, it is this indigenous capital-stretching capacity that is of the greatest importance, especially for the contemporary developing economy facing the formidable labour force explosion predicted for the 1970s and 1980s. It is in this specific area also that the scepticism of planners, engineers and aid officials generally is most pronounced, especially with respect to the range of technological choice which has been demonstrated to be really available when all the dust

---

[1] It is important to emphasise the word "efficient" since we are not concerned here with the (possibly quite legitimate) objective of employment creation as a separate social goal, to be traded off against output.

has settled. Using mostly historical examples from the Japanese case, we will now attempt to demonstrate the existence and potential importance of such capital-stretching innovations for the contemporary developing country.

LABOUR INTENSITY AND CAPITAL-STRETCHING IN THE FAR EAST

As has been pointed out by many observers, including Allen and Lockwood [1], the most significant feature of the Japanese landscape in the early Meiji period—following hard upon two centuries of self-imposed, nearly complete, isolation—was her ability to choose relatively freely from among the items on the technological shelf perfected in the West. The reopening of foreign trade and the resumption of other related contacts, especially the flow of technical personnel in both directions, led immediately to the stimulation of technological change by direct borrowing. But while the Japanese have often been characterised as possessing a consummate ability to copy and imitate, it is noteworthy that, in fact, the majority of domestic innovation activity very soon "consisted of the adaptation of foreign techniques to domestic conditions".[2]

The reasons for this relatively early move towards fitting industrial technology to domestic resource endowments are complicated and cannot be dealt with satisfactorily within the scope of this paper. Suffice it to say that post-Restoration Japan did not adopt very extensive or prolonged import substitution policies—partly because extra-territoriality deprived her of the ability to establish strong protective import barriers, and partly because the Government quite early thought it more efficient to work through the market (i.e. by using taxes and subsidies) rather than through extensive controls and government ownership. Those government plants in directly productive areas which were established during the immediate post-Restoration period were viewed mainly as pilot projects and were sold off to private interests around 1890. Thus Japan moved relatively quickly into the second phase of transition.

[1] George C. Allen: *Japanese Industrialisation: Its Recent Development and Present Conditions* (New York, Institute of Pacific Relations, 1940); W. W. Lockwood: *Economic Development of Japan, 1868-1938* (Princeton University Press, 1954).

[2] M. Miyamoto, Y. Sakudo and Y. Yasuba: "Economic Development in Pre-Industrial Japan, 1859-1894", in *Journal of Economic History* (New York), Vol. XXV, No. 4, Dec. 1965, p. 557. The same authors also report (p. 563) that similar capital- or land-stretching innovations took place during the same period in the agricultural sector, mainly through new cultivation methods on the intensive margin.

In assessing the importance of capital-stretching innovations (i.e. innovations which move the actual production shelf down to position S″ S″ in figure 1), it may be useful to distinguish between innovations relating to the machine proper, innovations relating to a given plant but peripheral to the machine, and innovations with respect to the production process as a whole, emphasising plant size and organisation at various stages of that process.

With respect to machine-related capital-stretching innovations, the simplest and quantitatively probably most important example was the running, at rates and speeds substantially in excess of those used abroad, of machinery imported from the United Kingdom and the United States. For example, once the kerosene lamp made nightwork possible, spinning could be done on two, sometimes three, shifts daily with but two or three rest days a month. This meant that the average work-week per machine was two to three times greater than that encountered in the country of origin; and, since physical depreciation is usually much less important than economic obsolescence, using a machine twice as intensively may not wear it out twice as fast. This heavy use of typical twentieth-century Japanese machinery meant that the normal gap between the physical and economic life of a machine was substantially narrowed and capital was considerably "stretched".

Moreover, there was a related speed-up of the very same spinning machines. By running the machines at faster speeds and/or by substituting cheaper raw materials (i.e. raw cotton—and making up for it by increasing the number of women to handle the resultant increase in the number of broken threads), an additional major saving in capital could be achieved:

> Certain differences in the industries of the two countries are important and must be noted. The raw material is essentially different. Though the Japanese do use some American raw cotton, the bulk of their cotton is from India and is of shorter staple, more likely to breakage . . . and requiring more labour to put it through the machinery. The yarn spun has more of the coarser counts that require more labour. . . . By adding more labour, [the spinning machinery] is run somewhat faster than American practice. . . . All of these factors are in some way related to the cheap labour policy. . . . They are there because the labour is cheap.[1]

Japanese spindles were equipped with a $\frac{7}{8}''$ instead of a $1''$ front roll to accommodate the shorter staple cotton when operated at higher speeds.

For these several reasons, namely differences in the yarn count and differences in the speed of the machine, as well as differences in the number of shifts, we find that there was a very marked substitution between capital and labour in the cotton spinning industry. For example, Orchard reports that a competent Japanese spinner working on a 20-yarn count operated

[1] John E. Orchard: *Japan's Economic Position* (New York, McGraw-Hill, 1930), p. 367.

from 300 to 400 spindles, while an American spinner on the same yarn count tended from 1,020 to 2,688 spindles, that is, between two-and-a-half and almost seven times as many.[1] As the United States Tariff Commission reported:

> In order to distribute the fixed overhead charges in the way of high interest and depreciation costs, and to earn the large amounts needed to pay a normal rate of dividend, every effort has been made to obtain the largest possible output from the expensive equipment and plant. Machinery is therefore run at high speed, and almost since their inception the Japanese spinning mills have been operated night and day, employing two 12-hour shifts (22 actual working hours) for an average of 27 days a month.[2]

Here again, given a standard count of yarn, the average Japanese spinner is seen as tending 240 spindles, while the American counterpart on the same machine tends about 1,000 spindles. As late as 1932 weekly man-hours per 1,000 homogeneous spindles of the same quality ranged from 328.8 in Japan to 164.8 in the United Kingdom and 143.1 in the United States.[3]

A somewhat similar story can be told with respect to cotton weaving:

> The high cost of mill construction is considerably reduced if you consider the hours during which the mill is being put to effective use. So far in Japan the wheels have turned round during 20 out of 24 hours, while in Europe only 8 hours are being worked. Effective working time in England is less than 38 hours per week, as 2 hours out of these are devoted to cleaning; this is done in Japan after working hours.[4]

Again, the United States Tariff Commission reports that "in weaving staple cotton sheetings, the ordinary Japanese weaver seldom operates more than two plain looms, while the American weaver, with perhaps some assistance in supplying fresh bobbins, normally tends from 8 to 10 plain looms".[5]

Perhaps the most convincing evidence that these adjustments along the machines proper constituted a rational response to very marked differences in factor endowments was the selectivity demonstrated with respect to the initial act of borrowing from the shelf. For instance, in weaving, in contrast to spinning, the latest automatic equipment from abroad was not, in fact, invariably imported. Quite frequently non-automatic looms were taken from the shelf instead, permitting more stretching than would

---

[1] John E. Orchard, op. cit., p. 367.

[2] United States Tariff Commission: *The Japanese Cotton Industry and Trade* (Washington, DC, Government Printing Office, 1921), p. 99.

[3] International Labour Office: *The World Textile Industry: Economic and Social Problems*, Studies and Reports, Series B, No. 27 (Geneva, 1937), Vol. I, p. 209.

[4] Arnold S. Pearse: *Japan and China* (Manchester, International Federation of Master Cotton Spinners' and Manufacturers' Associations, 1929), p. 86.

[5] United States Tariff Commission, op. cit., p. 100.

have been possible in the case of technologies to the left along that same shelf. Unlike some present-day less developed countries, Japan clearly did not wish to import ahead of its entrepreneurial and skilled labour capacities.[1] As the Tariff Commission put it:

> The price of the automatic loom is more than twice that of the plain loom, which, with the additional expense involved in the importation from the United States or Great Britain, made the total outlay too high in a country where the interest charges on money were relatively much higher than the cost of labour. Japanese mill managers have, therefore, hitherto preferred to employ more workers and to forgo the more labour-saving but more expensive machinery, in contrast to the situation in the United States where the high-priced labour is economised rather than the machinery.[2]

Taking cotton spinning and plain loom weaving on similar products together, the Commission reached the following conclusions:

> The average Japanese spinner or weaver tends about one-fourth the number of spindles or looms usually assigned to one person in an American mill. A comparison of the total number of persons employed in the two countries to operate individual plants of similar size, and, viewed more broadly, a comparison of the total number of persons employed in the whole American industry, per 1,000 spindles, with the number that would be required on the similar balanced basis under the Japanese conditions, confirms the general relation observed, that the Japanese mills require between three-and-a-half and four times as many operatives as the American.[3]

In the case of silk production, which, together with cotton, made up more than 70 per cent of total industrial output until the turn of the century, we have similar evidence of the ability to innovate in a capital-stretching direction on the machine proper. In raw silk, for example, the Japanese employed more than twice as many girls as did the reeling basins in Italy. In other areas, well into the twentieth century, Japanese railways employed 19 workers per mile of track compared with 7 in the United States.[4] In the production of printed goods, the following episode, which dates from the 1920s, may be instructive:

> Recently, a Japanese manufacturer of plain linoleum decided to undertake the production of printed goods. He dispatched a representative to the United

[1] United States Tariff Commission, op. cit., p. 116. The Commission reported that a shipment of automatic looms, imported shortly after the turn of the century, had been found so difficult to operate that the batteries and warp-stop motions were removed and they were run as plain looms, two looms to a weaver, instead.

[2] ibid., p. 116. A related interesting example of technical flexibility far beyond what most engineers are willing to admit to is provided by the Toyoda automatic loom, one of the few indigenous Japanese inventions in this area. Subsequently manufactured by Platt & Oldham under a Japanese patent, it was advertised to require 20 girls per loom in England, whereas 50 girls had always been used in Japan.

[3] ibid., p. 113.

[4] Orchard, op. cit., p. 375.

States to purchase the necessary equipment. The representative was familiar with the modern linoleum printing machine, printing several colours at one time and turning out as much as 15,000 square yards in 9 hours, but he considered it too expensive a piece of equipment, especially since his labour was being paid only about 50 cents a day, and so he sought out in an American plant an old hand block-printing outfit. It was not for sale. Its parts were lying about in a storeroom of the factory. Some of them were 40 years old, and the whole outfit had been discarded 15 years before. But the Japanese representative purchased it and had it shipped to Japan. In the immediate outlay of capital he saved money, for he purchased the old equipment at a price much below the price of a printing machine or even below the price of a new hand outfit, but he installed in his plant equipment that could only have been disposed of as junk in the United States. He started in Japan a new industry in a stage of technical development that had become obsolete years before in the older industrial countries.[1]

Many of the extra workers in Japanese plants are not engaged on the machine proper, but in what might be called machine-peripheral or handling activities. In place of mechanical conveyor belts, human conveyor belts are devised. Packaging and inspecting is more often done by hand. As Orchard again reports:

At one of the largest copper smelters in Japan, clay for the lining of the furnaces is carried down from a nearby hillside on the backs of women. At the plant of the Tokyo Gas Company, coke is put into bags by hand and then carried by coolies, some of them women, to the barges in the adjacent canal. Coal, even in the larger Tokyo plants, is unloaded by hand and carried in baskets to the power houses.[2]

The ability to substitute labour for capital in such activities peripheral to the machine proper apparently existed and its quantitative incidence was substantial. Very often such activities were machine-paced in the Hirschman sense, that is to say, while they might have looked wasteful to the untrained Western eye, they were in fact paced by well-spaced machinery as part of the same production line which contained large numbers of unskilled labourers.[3]

A third type of capital-stretching innovation of which much use was made in historical Japan is what might be called the plant-saving variety. This is often characterised by the coexistence of different historical stages of production in the same industry. Raw silk production and cotton weaving represent outstanding examples. In the former industry silkworm rearing and cocoon production were handled mainly by farmers' wives in small home-made sheds, extensions of the rural households. In cotton weaving, most of the yarn was "put out" to farm households, with individual looms dispersed

[1] Orchard, op. cit., p. 246.

[2] ibid., p. 255.

[3] This is very similar to contemporary methods of construction with the use of reinforced concrete in India and Pakistan. Here a cement mixer is linked to the final pouring of the concrete by a long chain of workers passing the cement from hand to hand; the cement is put in place just before it is ready to cool and harden.

in farmhouses and worksheds. But even in the more modern factory-style spinning industry, preparatory and finishing processes were carried out largely at the cottage level.

This rather remarkable survival of domestic industry on a subcontracting basis must be explained largely in terms of the exploitation of complementarities between many small labour-intensive operating units and the large industrial management unit. The traditional merchant middleman, as a representative of the subcontracting unit, served as both supplier and market for the goods to be worked up domestically. A specialisation of functions between workshops, even between the members of a given family, developed. One-roof economies could be achieved in this fashion (by using cheap labour in co-operation with old-fashioned machinery at the workshop level), while economies of scale could be achieved in the financing, purchasing and merchandising stages.[1] The continued relative importance of this household type of enterprise is quite remarkable; cottage-style industry contributed more than two-thirds of industrial output in 1878, and almost three-fifths in 1895, and retained substantial importance well into the twentieth century. Not only lacquerware, pottery, porcelain, saké, fruit and fish canning but also such new consumer goods as bicycles, electric lamps and rubber were to exhibit the same characteristics.

Plant amounts to more than 50 per cent of total investment in plant and equipment in most developing countries. The ability to utilise households for putting-out operations and thus reduce expenditures on plant undoubtedly amounted to a major kind of capital-stretching innovation. By deploying familiar but improving machinery over large numbers of scattered mini-plants, large amounts of unskilled labour could be deployed both in direct production and in satisfying the resulting increased demand for transportation and handling activities.

In this fashion, Japanese entrepreneurs were able to incorporate, first, pure labour services and, later, domestic ingenuity and skills, in the industrial production processes, largely for export. Other more recent examples of capital-stretching may be cited. In Taiwan, for example, after the liberalisation policies of the early 1960s substantially reduced distortions in the exchange rate, the interest rate and the intersectoral terms of trade, marked labour-using innovations took place in the textile, electronics and food processing industries. One example is large-scale mushroom and asparagus production as agricultural by-employment (similar to silk in Japan), combined with canning processes at the factory level utilising female labour with greater

[1] "Sometimes even a single part is not completed in one shop or home but is shaped in one and painted or plated in another." H. G. Aubrey: "Small Industry in Economic Development", in *Social Research* (New York), Sep. 1951.

intensity than anywhere else. Whereas traditional exports (mainly sugar and rice) still amounted to 76 per cent of total exports in 1955, by 1968 this had shrunk to 8 per cent of a much larger total. Meanwhile export substitution in the form of new agricultural products and, to an increasing extent, products embodying a large volume of pure labour services has taken hold. The ultimate expression of the latter trend can be found in the Kaohsiung Export Processing Zone, a tariff-free area into which, largely under subcontracting arrangements with Japanese or American firms, raw materials are imported and re-exported after value in the form mainly of unskilled labour has been added. Largely as a consequence of plant- and machinery-saving technological change of this type, Taiwan was, in 1970, reliably reported to be experiencing an unskilled labour shortage.

A similar trend has been in the making in the Republic of Korea. Devaluation and import liberalisation in 1963 and interest rate reform in 1964 laid the basis for major changes in the output mix as well as in the technology employed. In silk spinning, for example, 33 per cent more workers are reported to be employed per unit of capital than in contemporary Japan. A bonded export processing scheme, built on the same international subcontracting principle as that of Taiwan, now yields close to 20 per cent of an export volume which itself has been rising at an almost incredible annual rate of 30 to 40 per cent. In 1962 land-based foodstuffs and raw materials made up 75 per cent of total exports, while labour-based light manufacturing industries as a whole (including plywood, raw silk, cotton textile, wigs and footwear) amounted to 15 per cent. By 1968 the situation had been completely reversed, with 77 per cent of the exports in manufacturing and only 14.5 per cent in foodstuffs, livestock and raw materials. It should, moreover, be noted that small-scale firms manufacturing for export (i.e. employing fewer than 10 workers and undoubtedly the most labour-intensive part of the spectrum) grew from 18.6 per cent of the total in 1963 to 31.4 per cent in 1968.

In summary, the typical contemporary less developed country may be viewed as moving first through an import substitution phase in which pure technological transplantation is likely to be the order of the day, while shortages in domestic entrepreneurial capacity and other economic overheads are being repaired. Then, as the hothouse temperature is gradually reduced and the economy moves towards greater efficiency with the help of various liberalisation policies, labour-using types of technological change, both of the unintentional and of the intentional variety, assume increasing importance. In this phase the existence of the famous conflict between output and employment objectives in industrial development may be subject to fundamental challenge. Both the historical experience of Japan and that of the Republic of Korea in recent years illustrate that the current widespread scepticism concerning the supposed tyranny of the rigid technical coefficients may be seriously inaccurate. Such error derives in the main from an underestimate of the potential inventiveness of indigenous entrepreneurs, once they are given access, at a price, to the required inputs. And this is no trivial matter. For if our scepticism here is unwarranted, this would be among the most powerful arguments for accelerating the current, rather slow, trend towards liberalisation and the erosion of the substantial shadow price/market price differentials in factor and commodity markets.

DIESEL ENGINE MANUFACTURING: DE-AUTOMATION IN INDIA AND JAPAN

JACK BARANSON
International Bank for Reconstruction and Development

With evidence from diesel engine manufacturing, this paper demonstrates that adaptation of labour-intensive fabrication is not necessarily a feasible alternative to automation. Developing countries lack the labour skills to replace machine skills and the technical skills to adjust technology. Costs can run three to five times as high as the international standard. Once inefficient systems and practices are implanted, they are difficult to phase out or to convert to more efficient operations.

The ultimate logic of automation lies in its efficiency in the mass production of materials, components and end products. Modern manufacture and assembly require the putting together of a myriad bits and pieces that are interconnected, standardised and time-phased. These modern mass production systems depend upon science and technology to produce new materials, new products and new production techniques. Mass production also depends upon the conditioning and standardising of consumer tastes to assure high-volume production runs. Automation has emerged from three centuries of industrial evolution which have provided the metallurgies, the mechanical equipment and the power sources. This is because efficiencies depend upon exacting materials standards and manufacturing specifications upon tightly scheduled flows of materials and goods within and among plants and processing facilities.[1]

Automation, strictly speaking, is one step beyond mechanisation, which only replaces human energy and skill with machine power and sensitivity. Elevators and sewing machines are two examples of mechanisation. Automation involves electrical or mechanical devices linked to equipment that carry out predetermined tasks—a thermostat is a simple example of a device

[1] For a penetrating analysis of the anatomy of automation, see Robert A. Brady: *Organization, Automation and Society* (Berkeley, Cal., University of California Press, 1963).

that controls heating or cooling according to a programmed range of air temperatures. More complex automated equipment may roll steel to prescribed thicknesses or shape metal to tape-recorded dimensions. The modern ribbon-making machine for light-bulbs is a dramatic example of a self-feeding, self-operating, self-ejecting piece of automated equipment. Automation may be extended from an individual machine to an entire plant, as in cement making or oil refining; or segments of fabricating industries, such as body stamping and the welding of sub-assemblies (as in appliance and vehicle manufacture), may be partially or completely automated. The Volkswagen body plant in Wolfsburg (Federal Republic of Germany) is an example of a highly automated sub-assembly plant. Stamped body elements are fed into preset jig mountings and then simultaneously and automatically electrically welded. An entire car body is completed within an hour at some 25 stations on rotating and interconnected carousels. At certain volume thresholds, automated equipment and interlaced operations can turn out an expanding range of product variations. Consider, for example, the combinations of body styles, engines, accessories and trim offered in American cars.

Modern industrial systems differ from traditional artisan industries in several important ways. In mass production systems, every step and phase is conceived and perfected to minimise unit costs and standardise output. In an artisan craft, there is great heterogeneity in product design and individual pieces, and little thought is given to time-phasing and cost control. This basic difference often leads to difficulties in adapting artisan skills to factory labour. This was the case with an American diesel engine firm which experienced considerable difficulties with its plant in northern Scotland. Labour was recruited from among local craftsmen who were simply unwilling to yield individual skills developed over long years to the precision and standardisation built into "mindless machines".[1]

There are also essential differences in specialisation within and among production units. These differences are apparent if we compare the modern biscuit factory with a small bakery shop. The biscuit factory is an integrated complex of mechanical equipment and automatic controls, and working in conjunction with this factory are materials suppliers, equipment manufacturers, testing laboratories and distribution systems. The village bakery is a one-man show from the grinding of grain to the baking of the loaf; the baker may even fashion his own production utensils.

---

[1] The difficulties encountered by craftsmen in becoming "machine-minders" is poignantly described in Thorstein Veblen: *The Instinct of Workmanship and the State of Industrial Art* (1914). See also Gabriel Ardant: "Automation in Developing Countries", in *International Labour Review*, Vol. XC, No. 5, Nov. 1964, pp. 460 et seq.

ADJUSTMENT PROBLEMS IN LESS DEVELOPED COUNTRIES

Developing countries face some basic problems in attempting to minimise capital and foreign exchange expenditures for expensive automatic equipment or to maximise employment opportunities through the extensive utilisation of labour-intensive techniques.[1] High-volume automated equipment and techniques are also ill-suited to the limited market demands normally encountered in developing economies. Moreover, in protected "sellers' markets" there is not the compelling need for the cost efficiencies and rigorous quality standards associated with automation. Another factor inhibiting the use of automated equipment is the primitive stage of supplier industries upon which industrial plants depend for materials and parts.

But there are compelling reasons for choosing automated equipment despite volume and employment considerations. One is that quality and precision requirements in modern production systems place absolute limits on the technical feasibility of substituting human skills for machine capabilities. Several pieces of less automated equipment place a heavier burden upon supervisors' and operators' skills as regards adjusting tolerances, pacing the feeding of materials, and controlling the quality and reliability of finished parts. Sophisticated production may also require more backstopping and co-ordination. National deficiencies in technical and managerial skills may prompt factory managers to choose equipment that has precision and control built into the machine.[2]

The appropriateness of automated techniques depends therefore upon several interrelated criteria and constraints. These include scale production, precision or quality requirements, the stage of development of supplier industries, the availability of managerial and technical skills, relative cost and availability of labour to capital, employment policies, and foreign exchange resources. The case material which follows illustrates some of these difficulties and is drawn mainly from the author's work in the automotive and related metalworking industries.[3]

---

[1] For a survey of the literature on capital-intensive versus labour-intensive growth strategies, see Jack Baranson: *Technology for Underdeveloped Areas* (Oxford and New York, Pergamon Press, 1967), pp. 9-22. For an overview of decisions on automation, see Ardant, op. cit., pp. 432-71.

[2] See Albert O. Hirschman: *The Strategy of Economic Development* (New Haven, Conn., and London, Yale University Press, 1958), pp. 150-55, reference to machine-paced, process-oriented, capital-intensive techniques.

[3] See Jack Baranson: *Manufacturing Problems in India: The Cummins Diesel Experience* (Syracuse University Press, 1967); International Bank for Reconstruction and Development (IBRD): *Automotive Industries in Developing Countries*, by Jack Baranson, World Bank Occasional Papers, No. 8 (Baltimore, The Johns Hopkins Press, 1969).

KIRLOSKAR DIESEL ENGINES

An agreement to manufacture truck-type diesel engines was signed in 1962 between the Kirloskar Company of India and the Cummins Engine Company of the United States. The production plan called for the manufacture of 2,500 engines per year in India (about 2 per cent of the United States output) and reaching 90 per cent local content within about two years. The industrial product in this case was a highly sophisticated one in terms of manufacturing techniques, material requirements and quality standards necessary to produce a product commensurate with its economic costs. The 220-horsepower engine contained about 750 parts, ranging from cylinder blocks to fuel injection pins. A part such as a capscrew bolt requires about five processing steps; a part of middle-range complexity such as a crosshead valve requires 20 steps; and a more complex part such as an engine block takes up to 75 processing steps including castings, machining and finishing. This adds up to a total of about 15,000 separate processing steps and production techniques. In the United States about 40 per cent of these 750 parts are fabricated in over 200 supplier plants. In the manufacture of this Cummins engine, there are over 400 standards for materials (over 100 different varieties of iron and steel alone) and over 300 process standards and methods—all spelled out in over 3,000 pages of specifications.

At the United States plant in Columbus, Ind., equipment was designed and selected for high-volume capital-intensive production. Most of the machine tools were single-purpose, multi-station and multi-spindled. This meant that one machine might drill 60 holes in each of 5 cylinder blocks at a single pass. Narrow tolerances on diesel engine parts required the highest precision in tooling. Automatic control devices generally replaced operators' skills, thereby building quality control into the machine and reducing the need for inspection of machined parts. The more expensive quality machine tools are warranted only in high-volume production. Some multi-purpose single-operation equipment was used for low-volume parts at the United States Cummins plant, but even there, machine tools were tape-controlled for economy and precision. In converting equipment and techniques to the Indian plant's low-volume requirements, machine precision and control had to be replaced by human skills and organisational capabilities. Virtually all the extra tooling and fixtures required on the lower-volume equipment had to be custom made in the Indian plant's own tool-and-die shop. As a result, for a plant with one-twentieth the output, the Indian plant required a tool shop with about twice the facilities and three times the skilled labour of the United States plant.

A dramatic example of the scaling-down process, with its consequent upgrading of machine skill requirements and related factory logistics, was in the manufacture of crosshead valves. This small T-shaped part, measuring about 2 ″ by 3 ″, controls air intake and exhaust on each of the cylinders and requires a considerable amount of intricate machining and processing. The part requires about 22 steps to temper the metal, drill the pockets, thread the fittings, and weld a special alloy on to one surface. It is a moving part in the engine and must be manufactured to very close tolerances and specifications. In the United States plant, a single piece of automated equipment (a "Kingsbury") performed all machining operations and turned out three pieces every two minutes. Automated welding and hardening equipment completed the processing. In India at least 10 pieces of equipment along with special fixtures and jigs replaced the Kingsbury, and a hand-welding technique of detailed intricacy replaced the automated welding system used in the United States plant.

The following are further examples of "de-automation" which entailed additional training of machine operators and/or die-making skills to provide the necessary fixtures and jigs:

— in the finishing of fuel system plungers that fit into cone-shaped cups, hand-lapping (grinding and polishing) at the Poona plant replaced machine grinders normally used at the Columbus plant;

— at Poona hand-welding of tooth gears on to a ring base replaced automatic hobbing and milling from die castings at the Columbus plant;

— crankshafts at Columbus were milled from forgings that required expensive die castings, which were uneconomical for low-volume production. In India elements are milled and welded from steel bars;

— oil pans were manufactured in the United States from sheet steel moulded on a heavy-duty forging press, using permanent mould die-casting techniques. For low-volume production in India, sand moulds were used for iron castings which were machined on fitting surfaces. The sand-moulding technique was cheaper and industrially simpler but yielded a somewhat inferior product, as these techniques involve pouring cast iron; the resulting Indian product is thicker and heavier, and thereby adds undesirable bulk and weight to the engine;

— a sand-casting technique replaced castings from permanent mould dies on flywheel housings;

— iron castings replaced aluminium piston cylinders turned out at the Columbus plant. (Aluminium does not lend itself to sand-casting techniques but, again, iron added undesirable weight to the engine.)

In certain cases, relatively unskilled labour replaced automated devices. For example, manual labour in the Poona plant replaced various automatic devices used at the Columbus plant to convey materials, inspect components and parts, and assemble engines. These kinds of labour-for-capital substitutions were highly advantageous, since they did not involve the added burdens of training and technical co-ordination.

KOMATSU DIESEL ENGINES

A comparison of the Japanese and Indian experiences in adapting and absorbing imported technology highlights some of the technical conversion and manpower adjustment problems encountered by societies in early stages of industrialisation. At about the same time as the Kirloskar-Cummins venture was getting under way, the Cummins Engine Company also licensed a Japanese firm manufacturing construction equipment. Komatsu, the Japanese licensee, produced the same engine at the same low-volume level. But the Japanese firm was in a much more favourable position to convert technology and to handle the manpower adjustment problem. Within two short years they were producing the Cummins engine with 80 to 90 per cent local content and of a quality that equalled international standards and performance. In the Indian case it will take 15 years or more to come anywhere near the Japanese achievement.

Komatsu arranged for parts procurement on the above-mentioned crosshead valves from the Tsuzuki Manufacturing Company, a small parts manufacturer. A high degree of technical skill was required to convert techniques and produce the new technical drawings and manufacturing specifications. The special annealing process was broken down from automated techniques to hand-welding with detailed drawings and specifications to meet standards. The "Kingsbury" used in the United States was replaced in Japan by individual hand-machining operations. Hole tapping was done by a single operator, who sighted the stem and pocket centres, with apprentice machinists doing the final precision drilling. Tsuzuki suppliers used cheap lathes and depended upon human quality control. More refinishing had to be done in Japan on locally furnished forgings, which were of somewhat inferior quality at first. Technical drawings were prepared which gave detailed instructions concerning process steps, equipment requirements, millimetre dimensions and decimal tolerances. Specifications and tolerances include bevelling angles on the inside bore of the stem, eccentricity tolerances, treatment temperatures, cutting tools and machine speeds. Special instructions were given on avoiding cracking in grinding, keeping cutting tools free of metal powder and chips, and avoiding pin holes in stellite welding. The process sheet on annealing the stellite to the crosshead surface gave details on welding temperatures, distance and angle of flame to welding surface, and bubble and colour tests to assure correct welding temperatures. All this was for a single part that cost less than US$ 0.50!

The adaptation of "middle-range" machine techniques, which would be more advantageous to India's scale requirements and other factor endowments, were not as feasible as in Japan because of deficiencies in technical manpower and industrial organisation to convert technology and utilise less "intelligent" machines with higher labour skills. To begin with, Japan had an abundance of the experienced engineering and technical skills necessary to convert techniques to local equipment and materials. A second factor was the much higher level of machine labour skills and factory discipline developed in Japan over the past century. When advanced techniques with precision and uniformity built into the equipment were converted to labour-intensive production methods, heavier demands were placed upon machine operators to read blueprints, set up tools, and in other ways substitute human skill for machine accuracy. A third element found in Japan but lacking in India was an organisation of the industrial sector that permitted the effective use of small-scale shops as adjuncts to modern industrial complexes. To function effectively, these small shops had to be able to convert techniques to meet manufacturing specifications and, where necessary, adapt materials to meet standards. They also had to be able to co-ordinate their activities effectively and to schedule production within the larger industrial complex. India simply did not have the experienced and industrially disciplined small-scale industrial sector found in Japan.[1]

This ability of the Japanese to substitute human skills for machine capabilities or deficiencies in raw materials was already evident at the turn of the century. Second-hand textile machinery was imported by Japan from the United Kingdom, cheaper short staple cotton was used in combination with additional workers to mend broken threads, and repairmen were hired to keep older equipment going. Japan also now has an intricate network of small-scale shops and plants that are well integrated as suppliers into larger industrial complexes. These small shops are able to meet

[1] See also Toyoroku Ando: "Interrelations Between Large and Small Industrial Enterprises in Japan", in United Nations: *Industrialization and Productivity*, Bulletin 2 (New York, Mar. 1959), pp. 26-36. Smaller-scale factories and machine shops were able to compete with the larger, more modernised industrial plants by economising on capital costs and paying lower wage rates.

material standards and manufacturing specifications and to deliver goods on tight production schedules.[1]

OTHER COMPARISONS

An historical analogy may be drawn between India and the early efforts of the USSR to industrialise. David Granick, in his study of Soviet metal fabricating industries, points out that the advanced continuous flow techniques introduced by Soviet planners in the early 1930s were well beyond the organisational and technical skills then available in Soviet society, and that this resulted in widespread under-utilisation of over-capitalised industries. In order to be effective, the modern equipment needed industry-wide standards of quality control and efficiencies in interplant scheduling, which the Soviet economy did not approach until 30 years after the advanced techniques were first introduced.[2]

Manpower and organisational shortcomings among supplier industries were more formidable problems in India than in Japan. For one thing, the Indian economy did not provide the range of materials available in Japan, to say nothing of quality and standards. Many of the materials and parts procured in India were, at best, a near fit, or substandard in quality. Rejection rates on procured materials and parts ran anywhere from 10 to 15 per cent, as compared with 1 or 2 per cent in the United States. For example, in the United States engine-head bolts are made of a special carbonised steel and through-heated for hardening. In India they were made from the wrong steel and improperly heat-treated. As a result, the bolts snapped under tightening tension. There were also difficulties in procuring satisfactory castings for exhaust manifolds, thermostat housings and water pump connections; samples procured had a high porosity content, which resulted in leaking parts. Radiators and fans meeting Cummins specifications were two other items difficult to procure locally. Other rejection items included: filter cloths that failed strength tests; rubber liners and sealing rings with surface defects and non-pliability at low temperatures and insufficient resistance to oil; compression rings that were too brittle

---

[1] See Gustav Ranis: "Factor Proportions in Japanese Economic Development", in *The American Economic Review* (Menasha, Wis.), Vol. XLVII, No. 5, Sep. 1957, pp. 594-607; and United Nations Centre for Industrial Development: "The Dual Nature of Industrial Development in Japan", in United Nations: *Industrialization and Productivity*, Bulletin 8, (New York, 1965), pp. 41-52.

[2] David Granick: *Soviet Metal Fabricating and Economic Development: Practice versus Policy* (Madison, University of Wisconsin Press, 1967). See also review of Granick's book by Jack Baranson in *The American Economic Review*, Vol. LVIII, No. 4, Sep. 1968, pp. 1028-29.

or insufficiently hard; steel bearing strips that did not bond properly; bearing caps with objectionable graphite flaking; cylinder liners that failed hardness and tensile strength tests; and copper gaskets that were too hard.

Another manpower and organisational deficiency in India related to formulating and carrying out a production programme. Automated systems depend upon tight production controls, which took several years to introduce into the Poona plant at anywhere near satisfactory performance levels. For example, production controls require a detailed and exhaustive list of the parts to be manufactured, from which process sheets detailing machine loads and tooling requirements (machines, equipment, fixtures and gauges) are normally prepared. It is then necessary to co-ordinate and time-phase all departments in support of manufacturing operations: purchasing (to procure the necessary parts and materials), quality control (to approve purchased raw materials and semi-finished parts), production control (to schedule the flow of materials and parts), and manufacturing engineering (to prepare the list of required machines and tooling). When these procedures are not followed in a systematic and comprehensive fashion, the inevitable results are shortages, bottlenecks and much idle equipment—all of which defeats the basic purpose of continuous flow techniques in the modern industrial factory.

India's substantial deficiencies in industrial management also proved to be formidable barriers to the effective utilisation of plant equipment. Lead times to procure equipment and fixtures were either underestimated or ignored in India. Since no check had been kept on the status of procurement, several pieces of essential equipment were not received on time. Either domestic delivery had been held up, or import licences had not been cleared. Other equipment to machine essential parts had not been ordered: a milling machine for engine blocks, a hobbing machine and gear grinder for gears, and an induction hardening machine for rocker levers and crosshead valves. As a result, there was an 18-month delay in the machining of engine blocks and parts. (Part of the difficulty was also the low grade of sample castings for blocks and most other major parts.)

Economic conditions endemic to developing economies and prevailing in India placed an added burden upon industrial management in that country and undermined the effective use of automated equipment, which must be used to full capacity if it is to be economic. Erratic fluctuations in effective demand in India also seriously undermined production programmes. In the Kirloskar-Cummins case potential customers from among original equipment manufactures had been carefully surveyed, but by the time the plant was in full operation actual orders amounted to only 500 engines for a plant equipped to produce 2,500 on a two-shift basis.

In the two critical areas of production controls and quality standards, the Indian management's philosophy was one of "doing its best" under adverse conditions. Bottlenecks and inadequacies were dealt with as they appeared, rather than by laying down and then adhering to carefully conceived production plans. Indian managers viewed the insistence by their American partners on rigorous planning and strict accounting procedures as either unreasonable or inapplicable to the Indian situation. Ultimately, the Indian management was quite satisfied to utilise expensive and sophisticated equipment to turn out simple parts for a cheap line of small diesel engines. Such views are simply incompatible with industrial systems to produce standardised quality products using automated techniques. Indian philosophy and practice was a far cry from industrial performance by the Japanese licensee, where control systems were religiously established and rigorously implemented. Komatsu, the Japanese partner, engaged in an intensive interchange with the Columbus management over specifications, details on process sheets, and adjustments to slight variations in local materials or tools.[1] There was virtually no such interchange between Poona and Columbus.

BROADER PROBLEMS OF AUTOMOTIVE DEVELOPMENT

The described difficulties in establishing diesel engine manufacturing in India represent a microcosm of the broader problems now faced by more than 30 developing economies that have embarked upon progressive production of automotive vehicles. What has been said previously about diesel engine manufacture (the heterogeneity of materials, rigid materials specifications and manufacturing standards, and the high engineering and managerial skills required to ensure the necessary quality and reliability) applies to the fuller range of automotive component and parts production. In developed countries, high-volume automated techniques are used to manufacture all but a limited range of specialised vehicles and parts. Automated transfer

---

[1] Written inquiries over a short period included questions on how to agitate blasting compound, deburr parts surfaces, sharpen milling cutters to get correct parts finish, apply and cool special stellite welded on to a crosshead part, and regrind gears after heat treatment. Highly detailed instructions were then issued by the Cummins plant on how to wash and deburr parts to improve quality, minimise excessive milling on cylinder blocks and other heavy parts, hold tolerances on the depth of drill bores, use air jets to keep tools free of metal chips during cutting cycles, reduce burning-up of bits on drilling operations, improve grinding wheels to cut down "burning" of parts surfaces, use a new type of polishing cloth to remove grinding marks, adapt an indexing mechanism on a boring machine, avoid distortion of gear teeth after heat treatment, and reduce wear on cutting tools used to machine parts after metal hardening. See Baranson: *Manufacturing Problems in India*, op. cit., p. 45.

lines (including rearrangeable standard machine elements) reduce operating and handling costs, increase the rate of utilisation of expensive equipment, and reduce costs for machine tools, factory space, rejected parts and machine maintenance.

This fragmentation of production, which has characterised automotive industries in developing countries, inhibits the efficient use of automated techniques. But the demands for precision and standardisation require automated equipment far beyond what scale considerations alone would allow. This has resulted in widespread over-capitalisation. For example, in 1967 the Latin American Free Trade Area, as a result of national development along the lines outlined above, had 10 times the plant and equipment required to manufacture the total 650,000 vehicles in demand. The region, incidentally, had more than 60 firms producing at least 200 different models and makes, where from an economic standpoint three firms producing no more than half a dozen major lines would be justifiable.[1]

Extensive import substitution, both horizontally into diversified products and backward vertically into supplier industries, has contributed to widespread shortages of managerial and supervisory personnel to implant production systems and related controls. It takes 20 to 30 technical supervisors and managers on assignments ranging from three months to five years to get a plant operating in a country like Brazil. The case of diesel engine manufacture in India indicated some of the critical areas, such as production engineering and quality control, where less developed countries are especially deficient. These skills are highly important in production for world markets where quality standards and cost efficiencies are much more demanding than in protected sellers' markets.

Various approaches may be followed to the challenges posed by automation. In certain instances, restructuring the industries producing for the home market may lead to the higher production volumes needed to make advantageous use of automated techniques. Rationalisation in the automotive industry would involve the drastic reduction in the number of models and makes that are now generally produced in developing countries. In certain instances, programmes to standardise major components such as engines and transmission may prove feasible. Other segments of automotive production which lend themselves to labour-intensive techniques include body fabrication of buses and large truck chassis, interior trim on passenger cars, electrical wiring, and most assembly operations. Stabilisation of design cycles over extended periods as was done by Volkswagen, Volvo,

[1] See Jack Baranson: "Integrated Automobiles for Latin America?", in *Finance and Development* (Washington, DC), Vol. 5, No. 4, Dec. 1968, pp. 25-29.

Citroën and most other European motor car producers, is another means for attaining sufficiently large production runs. Modular design of body panels and the use of standard steel shapes for body and chassis elements are other means of achieving volume. Several years ago, United States Steel designed a series of vehicles (passenger car, farm tractor, and small utility truck) based upon 20 to 30 interchangeable body elements. In certain cases, shifts to other materials and techniques may permit a moderate degree of automation. This has been the case in the use of fibreglass car bodies, which can be produced efficiently in much smaller series than metal bodies.

In some instances, moderate adjustments in product designs and production techniques would be of considerable advantage in arriving at a proper fit between acquired technology and emerging stages of industrial capability. The fact is that, despite considerable differences in production environments (and consumer needs), changes are rarely made in product design or basic production techniques (other than scaling down plants). In developing countries a case may be made for more functional design that fits conditions of demand and supply for industrial products. This does not mean that a sophisticated product such as a high-speed diesel engine should be manufactured at high costs or of poor quality or performance. What is needed is industrial design to adjust the product to functional needs and productive capabilities so that the product may be manufactured efficiently. Developing countries would benefit greatly from discrete development of indigenous research and engineering capability at the industrial plant levels to help bridge this technological gap.[1]

Much can be achieved in the way of intelligent choices and adaptations of production methods through "work-restructuring" analysis, which can help improve productivity of man/machine relationships through the reorganisation of tasks and control patterns.[2] Assembly-line operations have been restructured into product-oriented work groups and individual tasks enlarged or enriched to reduce the monotony of mass production and thereby improve productivity. Such groups may combine tool set-up and minor maintenance, assembly or processing, and inspection—instead of routinised, short-cycled, repetitive tasks. These techniques are now used extensively by modern industrial organisations and can be adapted to the particular cultural contexts and levels of skills in a developing country.

---

[1] See Jack Baranson: "Role of Science and Technology in Advancing Development of Newly Industrialising States", in *Socio-Economic Planning Sciences* (Oxford and New York, Pergamon Press), Vol. 3, No. 4, Dec. 1969, pp. 351-83.

[2] See NV Philips Gloeilampenfabrieken: "Work-Structuring: A Summary of Experiments at Philips, 1963 to 1968" (Eindhoven, 1969); and William J. Paul *et al.*, "Job Enrichment Pays Off", in *Harvard Business Review* (Mar.-Apr. 1969), pp. 61-78.

Work restructuring techniques ultimately can be carried back to the technical design phase of product and production engineering. In some cases, minor adjustments in product design may also permit work restructuring which is more in line with the skills and motivational patterns of a particular society.

Export markets in specialised products and components provide major opportunities for utilising selective segments of automated techniques. This is particularly true where low wages can be applied productively and where production programmes permit the import of materials and components that cannot be produced competitively in the local economy. It is also necessary that foreign technical and managerial personnel be allowed into the country until indigenous manpower becomes sufficiently experienced to take over. In the automotive field, export possibilities might include: (*a*) the manufacture of specialised components and parts; (*b*) responsibility for a particular vehicle line; (*c*) specialisation in low-volume replacement parts for obsolete models; or (*d*) the reconditioning of engines and parts for world markets.

Several agreements have actually been signed in some of the areas outlined above. One is for parts export from Mexico by Massey-Ferguson. Another involves Fiat, which is now going ahead on plans to transfer an entire passenger car series to its affiliate in Yugoslavia. Colombia and Mexico have revised their automotive decrees in order to encourage foreign partners of automotive manufacturers to produce or procure parts for export. Under these new decrees exports can be developed instead of a much broader range of components being duplicated to meet local-content requirements.[1]

Under proper marketing and managerial organisation, middle-range industrial skills may be used effectively in fairly sophisticated activities involving moderate degrees of automation. The electronics industry in Taiwan is a good case in point. In 1970 more than a dozen plants were manufacturing, for export, sub-assemblies for radio and television sets and related electronic devices and were carrying out very sophisticated manufacturing of micro-electronic devices. A typical plant is staffed by 15 foreign technicians and employs over 1,000 people. In the manufacture of micro-electronic devices, most of the exacting precision standards are built into specialised processing equipment such as laser guns for welding under microscopic sights. This is matched by factory skills that do the delicate welding and assembly work and are trained in a matter of months. In

---

[1] It should be pointed out that overvalued exchange rates and indiscriminate import substitution do price domestic supplier industries out of world markets. This means that, to be significant, export efforts require some basic policy changes in protectionism. See Harry G. Johnson: *Economic Policies Toward Less Developed Countries* (Washington, DC, The Brookings Institution, 1967), p. 245.

the electronics plants of Taiwan one can see the transition from hand skills to advanced industrial systems.

Another less spectacular example of this kind of opportunity was the development of aircraft maintenance and overhaul facilities in Costa Rica as an auxiliary (SALA) to the national airline. SALA was certified by the United States Federal Aviation Authority to perform a standard six-month overhaul on certain classes of aircraft equipment (including passenger services); this encompassed complete engine and navigational system breakdown. SALA had overhaul contracts with air carriers such as KLM and Pan American.

In considering export markets it should be pointed out that automated techniques may be as warranted in manufacturing biscuits as in fabricating engine parts. The conventional wisdom on the choice of production techniques associates labour-intensiveness with simple consumer goods. But under competitive conditions, where standardisation and volume are important, varying degrees of automation may be justified. One such case is the Minoo factory in Iran which, in addition to manufacturing a broad range of biscuits and confectionery, also produces pharmaceuticals and cosmetics. These latter activities were an outgrowth of the fairly sophisticated laboratory control and packaging facilities built in connection with the sweets and biscuit operation. The Minoo plant has one of the largest continuous-baking ovens in the world. The sweet-making equipment is also highly automated, including mixing, extruding and wrapping. The plant does make judicious use of manual labour for final packaging and sorting. Substantial laboratory facilities were built to develop suitable tastes and textures for the national and regional market and to quality control production. Despite the relatively high cost of the protected raw sugar that Minoo must use, a small percentage of the plant's output is now exported to Persian Gulf countries and to Afghanistan in competition with European producers.

International firms are in a particularly favourable position to assist their manufacturing affiliates in developing countries to specialise for world markets. These firms have the technical resources and the access to world markets which developing countries must acquire if they are to export. Foreign enterprises are also in a position to assist developing economies in adjusting imported technology and developing long-term research and engineering capabilities. In some cases they can help redesign product groups that are more in line with technological absorptive capabilities. This may mean adjustments in the degree of automation or in the product itself to accommodate high-volume equipment. But most important, the multi-national firm's access to world markets provides the best opportunities for using automated techniques in volume production.[1] The operations of Fiat in Yugoslavia and Massey-Ferguson in Mexico are cases in point.

Frequent reference has been made in this paper to the dependence of automated systems on scheduling and control. Foreign enterprises can also assist manufacturing affiliates in developing countries to upgrade management systems and skills. Volkswagen, for example, has put considerable effort into management guides which set forth procedures in great detail on everything from quality control to service department organisation. By so doing, they hope to accomplish two aims: to decrease the need for management skills at the receiving end, and to reduce the number of critical personnel engaged in transferring production systems.[2]

Finally, more realistic policies for dealing with unemployment problems would open the way for more effective utilisation of industrial manpower and automated techniques. Instead of trying to create jobs extensively throughout industry by the establishment of plants of uneconomic size and the over-use of labour-intensive techniques, industrialisation programmes should be formulated to throw the burden of employment creation on sectors where relatively large amounts of labour can be used economically. This includes a wide variety of other secondary and tertiary industries. In some sectors (for instance, food processing) an advanced marketing and production system, including automated techniques, can provide the leading edge for growth and massive employment in supporting sectors (i.e. food growing). One cannot separate employment problems and choice of technique from economic policy in general.

[1] See James Brian Quinn: *Scientific and Technical Strategy at the National Enterprise Level*, paper presented to UNESCO Conference on the Role of Science and Technology in Economic Development, Paris, 11-18 Dec. 1968. See also Baranson: "Role of Science and Technology . . .", op. cit., pp. 363-65.

[2] See Werner P. Schmidt: *The International Transfer of Management Skills—Volkswagen's Needs, Experiences and Plans*, paper presented to AIESEC (Association internationale d'étudiants de sciences économiques et commerciales), Turin, 19 Nov. 1969.

TECHNOLOGY AND ORGANISATIONAL REQUIREMENTS IN DEVELOPING ECONOMIES

PETER KILBY
Wesleyan University, Middletown, Conn.

This paper attempts to assess the kinds of organisational requirements and specific technological skills that are associated with the introduction of more advanced technologies in underdeveloped countries. The paper is in three sections. The first presents a general theoretical framework for interpreting the nature of technological advance and concomitant skill requirements; the special case of automation and its applicability in low-level technology economies is examined. In the second section, work competency requirements and actual performance are examined in three industries. The third section addresses the issue of manpower training in the light of the material presented in the first two sections.

ECONOMIC DEVELOPMENT AND TECHNOLOGICAL CHANGE

The process of economic development—roughly measured by a rise in national income per head from somewhere in the neighbourhood of US$ 100 to US$ 500 and above—can be examined from many viewpoints. The economist typically analyses the process in terms of: (*a*) aggregate capital formation and sectoral investment patterns; (*b*) the development of an efficient, interlocking network of commodity, factor and money markets. The economist recognises thrift, technology and tastes as critical variables, but he treats them as essentially exogenous. The sociologist studies the patterns of role differentiation and integration in social structure which parallel economic specialisation and market co-ordination. The social psychologist concentrates on the kinds of changes in the individual's attitudes and motivations consonant with the pattern of modern activities. Each of these perspectives uncovers important causal factors, both permissive and catalytic, in the complex and variegated modernisation phenomenon.

Technology and organisational requirements

It is possible and, for the purpose of this paper, quite useful to view economic development as a process of technological change with its attendant changes in organisational and skill requirements. We shall start with a description of the organisation of work in the traditional economy and then sketch in the direction of entrained changes that coincide with the introduction of progressively more productive technologies.

In traditional society, technology (both agricultural and industrial) is non-complex. The production unit (almost always the family) is small, and capital equipment is limited to hand-wielded and animal-drawn implements with few or no moving parts. Production is solely for the consumption of the producing unit. Raw material input, product specification and product quality are variable. Skills are comparatively simple and are passed on from generation to generation through observation and learning-by-doing. Learning based on comprehending general principles and analysis is therefore minimal.

With the exception of a few capital goods that are produced at infrequent intervals, traditional technology is characterised by an absence of intra-commodity specialisation of labour. The hoe, the pot, the cloth are produced sequentially by an individual moving from one operation to the next until the product is completed. With no intracommodity division of labour there is no call for the co-ordination of simultaneous operations, supervision or controlling production flow. In the case of group-produced capital goods, such as a canoe or a dam, co-ordination is supplied by a chant or some other ritualised ceremony rather than the independent initiative of a "manager". In short, with each individual producing a commodity in its entirety at a pace and quality of his own setting, no internal management functions are required. Moreover, with each unit producing for its own consumption, no external management functions associated with producing for market sale are called for.

As more productive technologies are introduced in non-agricultural production, increasing division of labour and market interdependence occur at every level. (The seasonally sequential nature of farming precludes the necessity of task specialisation.[1]) Productive units specialise in the manufacture of a single commodity or a part of one commodity. Workers within the enterprise specialise in one component activity of the production process, so well described by Adam Smith's pin-making example. Breaking down the production process into its constituent parts facilitates the introduction of machinery which leads to standardisation and greatly multiplied

[1] John M. Brewster: "The Machine Process in Agriculture and Industry", in *Journal of Farm Economics* (Menasha, Wis.), Vol. XXXII, No. 1, Feb. 1950, pp. 69-81.

productivity of human effort. The more complex technologies require that the human organisation should command a higher level of knowledge and a wider range of specialised skills, as well as the more advanced technology embodied in the capital equipment which it employs. With the emergence of specialisation, internal managerial functions must be performed to synchronise and co-ordinate each subset of activity and to control for quality, standardisation, materials wastage and rate of through-put.

Intracommodity division of labour entails economic specialisation, with self-sufficiency being abandoned for market dependence. Division of labour increases the scale of production, which in turn necessitates that output must be marketed and factor inputs must be purchased. Market-oriented management functions must now be performed to sell the product and selectively purchase a wide diversity of capital equipment, raw material and labour service inputs. There are labour relations, customer relations, supplier relations and public bureaucracy relations to be managed. The financing of circulating and fixed capital must be secured and maintained.

Beyond its impact on the type of technological and organisational skills required, the process of moving to more productive, specialised technologies entails major changes within the production unit and in the over-all structure of producing establishments in the economy. Individual production units grow in scale, employing a larger capital stock and more capital per employee. Superior technology means that both capital cost and labour cost per unit of output fall, although capital cost typically falls less. The rising capital/labour ratio occurs because of the nature of the more productive technologies available for transfer from the capital-rich economies, and because of the tendency to substitute capital for labour as wage rates rise faster than interest rates in the home economy. As division of labour proceeds within the firm, those processes that are subject to increasing returns to scale are taken over by specialist firms which, by obtaining a scale of production that permits full cost minimisation, can then sell this input to a number of firms at the previous stage of production cheaper than any one of these firms can themselves supply the input.[1] Thus value added per unit of gross output falls for the individual firm, and new firms in vertically linked industries emerge. This process of growth continues until scarcity of inputs, limitations of final demand or the exhaustion of economies of specialisation bring the process to a halt.

How do manpower requirements change with successively more complex technologies? We have classified skilled work performance in the following three categories:

[1] George J. Stigler: "The Division of Labor is Limited by the Extent of the Market", in *Journal of Political Economy* (Chicago), Vol. LIX, No. 3, June 1951, pp. 185-93.

1. Specific technological skills.

2. Internal production management functions: (*a*) planning and co-ordination; (*b*) supervision; (*c*) control of quality and input utilisation.

3. Market-oriented management functions: (*a*) adapting to market opportunities; (*b*) marketing of output and purchasing of inputs; (*c*) human relations management—with employees, customers, suppliers and the public bureaucracy; (*d*) financial management.

We have noted that in the traditional economy only the first category of skills is present. With specialisation and production for sale, categories 2 and 3 are called into play. Category 3 differs from 1 and 2 in that market-oriented management functions are not significantly altered with changes in technology; rather the intensity of these functions tends to vary with the size of the enterprise. In large contemporary firms, data may even be processed electronically; nevertheless, a sales manager or personnel manager for a given size of firm in 1970 is involved in activities that would be wholly recognisable to his counterpart in 1870. But an aircraft engine mechanic of 1920 would have as little competence to repair and maintain a 1970 jet engine as a book-keeper of 1940 would have to operate a key punch.

When one speaks of manpower requirements and education, one is usually referring to the specific skill competencies of category 1. When reference is made to actual employment positions (technician, engineering assistant, foreman, etc.), a mixture of categories 1 and 2 is involved. The frequent assertion that underdeveloped countries have a great need for intermediate manpower skills of the type supplied by formal vocational training institutions usually turns out to be incorrect. Careful analysis of wage-rate data, employer recruitment practices for supervisory positions and the experience of technical education institutions reveals that the actual need is for supervisory skills and not for the higher-level technical skills. Failure to observe this distinction was precisely the mistake of the Ashby Commission and subsequently the National Manpower Board in Nigeria, which has resulted in a vast over-expansion in the number of technical colleges in that country.[1]

The tripartite categorisation of skilled work activities we have been employing is also useful for analysing training requirements. Experience has shown that with the appropriate arrangements (a thorough grounding in the underlying physical principles, applied training in the workplace and the proper motivational setting) there are no serious impediments to the

[1] Peter Kilby: *Industrialisation in an Open Economy: Nigeria 1945-1966* (Cambridge University Press, 1969), p. 254.

creation of modern technological skills in the developing countries. Similarly, long traditions of exchange relationships and "political administration" in semi-traditional societies appear to have provided sufficient antecedents, with one exception, for the successful performance of market-oriented management functions.

The area where the most persistent performance shortcomings are to be observed, as documented by the writer [1] and W. Paul Strassmann [2] among others, lies in the domain of internal co-ordination, supervision and control. One explanation is that the comparative absence of intra-commodity specialisation in traditional technology has precluded any antecedent roles for these functions. Further, the writer would argue that social structural factors (the plurality of purpose of traditional labour, wherein economic services are intertwined with other basic social functions, and the existence of hierarchical status systems) operate to impede effective supervisory activities. Authoritarian social structure can also be invoked to explain the one exception in category 3, namely difficulties in labour relations, so often described for India and Latin America.[3] We shall return to the training aspects of category 2 in the third section.

In examining the manpower aspect of advanced technology, perhaps the most useful approach is the categorisation of production techniques as process-centred or product-centred made popular by A. O. Hirschman.[4] In the first type, the technology consists of a basic chemical or mechanical process; all other operations are comparatively simple and are powerfully entrained and specified by the central process. Examples are smelting, petroleum refining, fertiliser, cement and beer. In product-centred industries there is no single process which accounts for the bulk of the transformation, with the result that the pace and quality of manufacturing operations is far less rigidly compelled. Furniture, agricultural equipment, construction and metalworking are a few examples.

Hirschman notes that where product-centred industries can organise a flow (as on, for instance, a conveyor-belt assembly line) it may be able to imitate the conditions of a process-centred industry. He proposes that an industry can be placed into one or other of these categories by asking whether a production unit can be assigned a definite capacity. If capacity can be given, it is process-centred. "In the product-centred industries as,

---

[1] Peter Kilby: "Hunting the Heffalump", in *Entrepreneurship and Economic Development* (New York, The Free Press, 1970).

[2] W. Paul Strassmann: *Technological Change and Economic Development* (Ithaca, NY, Cornell University Press, 1968).

[3] Everett E. Hagen: *On the Theory of Social Change* (Homewood, Ill., Dorsey Press, 1962).

[4] Albert O. Hirschman: *The Strategy of Economic Development* (New Haven, Conn., and London, Yale University Press, 1958), pp. 147-48.

for example, in the case of a construction firm or a repair shop, it is not possible to make this kind of statement: outputs here are often heterogeneous and even if the problem of adding them up can be satisfactorily solved, capacity is far less rigidly set by the physical assets alone or it is a far less useful benchmark because actual output seldom reaches more than a fraction of theoretical capacity." [1] Hirschman's argument is not wholly convincing in that time-and-motion studies are explicitly or implicitly made in all product-centred industries. Construction and repair firms quote specific prices for specific jobs, thereby implying a rate-of-operation norm. In 1969 the writer visited a number of small manufacturers and assemblers of two-wheeled tractors in Taichung, and typically the managers, who were modestly educated and possessed no formal industrial training, employed a notion of 100 per cent efficiency against which they compared actual output.

Even with these qualifications, it remains true that process-centred technologies substantially reduce managerial skill requirements. They provide a work-pace, a capacity maximum against which to measure actual performance, control over materials input utilisation and a brake upon variation in quality. A fully automated plant represents the extreme case of the machine-paced, process-centred production process: it removes the need completely for direct production labour and all supervisory and managerial services other than those required at the first and last stage of production. In many intermediate cases, such as the use of computers in banking or record-keeping departments, automation is labour-displacing and cost-reducing; while it economises on superintendence in those areas where it is applied, it does not necessarily have a significant influence on managerial efficiency elsewhere.

Because automated techniques are, inter alia, a substitute for an underdeveloped economy's scarcest commodities (supervisory and controlling competencies), it is frequently stated that automated processes should be used as widely as possible in these countries. Specifically it is argued by Hirschman and others that the efficiency gap between producers in advanced and backward economies will be far smaller for machine-paced than it will be for man-paced industries. Thus a fertiliser plant in India will operate within, say, 5 or 10 per cent of the utilisation level of a similar plant in the United States; on the other hand, an American diesel engine manufacturer in India utilising a product-centred process operated at 20 per cent of capacity as compared with about 85 per cent for its home plant.[2]

---

[1] Hirschman, op. cit., p. 148.

[2] Jack Baranson: *Manufacturing Problems in India: The Cummins Diesel Experience* (Syracuse University Press, 1967). For further information on this case, see also Dr. Baranson's paper in this volume.

Such a favourable evaluation of the applicability of automated technology to developing areas is (because it focuses on but a single factor) quite misleading. There are two other considerations that impair its suitability. First, automated technology is capital-intensive, and the services of capital are substantially more expensive in developing economies because: (*a*) interest rates are higher than in capital-abundant countries; (*b*) the capital equipment usually must be imported—for example, the installed cost of a fertiliser plant in India is 140 per cent of that in the United States and Europe.[1] Second, automated technologies require a uniform and uninterrupted supply of power, raw materials, processed inputs, specialised maintenance services, and so on, the reliability of which tends to be a function of the economy's level of development. As an example, the rejection rate for locally purchased component parts by the Cummins diesel engine factory in Poona described by Baranson runs at 11 to 21 per cent as compared with 3 to 4 per cent for its Columbus plant.[2] In short, unless automated production is associated with some natural resource advantage (e.g. natural gas for fertiliser) or high transport protection (e.g. inland cement factories), or untraded commodities (e.g. electricity, banking services), it is likely to represent an uneconomic use of resources.

THREE INDUSTRIES

In order to obtain a clearer picture of specific manpower bottlenecks associated with the introduction of advanced technologies, we now turn from general considerations to an examination of industrial efficiency and work performance in three industries: agricultural implements, motor vehicles, and ammonia-based fertilisers. In the final section we will discuss the kinds of educational arrangements that are best suited to meet these needs.

Agricultural implements

This is a multi-product industry based upon a metalworking, product-oriented technology of low to intermediate complexity. Information was collected by the writer and Professor B. F. Johnston on visits to about 30 producers [3] in India, Pakistan and Taiwan during August and September 1969.

---

[1] Organisation for Economic Co-operation and Development (OECD): *Supply and Demand Prospects for Fertilisers in Developing Countries* (Paris, 1968), p. 84.

[2] Baranson, op. cit., p. 72.

[3] Products included carts, waterpumps, tubing and coir strainers, diesel engines, electric motors, maize shellers, stationary threshers, pedal threshers, hoes and shovels,

Production processes are quite numerous, although fairly simple. Iron components are sand-cast from pig-iron ingots and scrap. Steel parts are cut from imported rods or sheet. Wooden elements are also cut and shaped in the factory. All components are machined and then assembled and tested. Equipment includes casting furnaces and sand forms, lathes, drill presses, woodworking machines, and hand tools. In all the smaller establishments (i.e. fewer than 50 employees) production is by batch; in the larger firms, it is on a semi-continuous flow basis.

Compared with the motor vehicle industry, tolerance for quality variation in agricultural implements is quite high. Thus there are noticeable variations in product specifications and in the quality of casting and finishing. Only one of the firms with fewer than 50 employees is engaged in producing power tillers. Labour in these establishments seldom has any formal education beyond primary school; technical training is by unregulated apprenticeship. The educational background of manager-entrepreneurs ranges from middle school to bachelor's degree. Written records do not appear to be used extensively as a control or costing technique.

In the larger establishments, which produce tractors and power tillers, both quality control and the skills of the labour force are very much higher. These firms employ graduate engineers (about 3 per cent of the labour force) as well as many qualified technicians. The bulk of their labour force is recruited from vehicle repair firms and vocational schools. Specialised machinery is used to test the tolerance of critical parts. Some 40 per cent of all components are produced by small ancillary firms; another 25 to 30 per cent (typically the gearbox, crankshaft and fuel pump) are imported. Full-blown record-keeping and control systems are utilised in these factories. Because of interrupted supplies of one component or another, the flow organisation of production is not fully attained. Under-utilisation of productive facilities is substantial.

Motor vehicles

The works of Jack Baranson provide considerable data on our second case, the motor vehicle industry.[1] Closely related to the tractor case, this

---

cotton gins, baling presses, levellers, harrows, rotary weeders, sweet-potato slicers, sprayers, tractors, power tillers, grain driers, ploughs, cultivators, row markers, chaff cutters, seed drills, sugar-cane crushers, inverting ploughs, low-lift water pumps, diesel-powered pump sets, seed planters, and tractor-drawn implements.

[1] International Bank for Reconstruction and Development: *Automotive Industries in Developing Countries*, by Jack Baranson, World Bank Occasional Papers, No. 8 (Baltimore, The Johns Hopkins Press, 1969); Baranson, *Manufacturing Problems in India*, op. cit.; idem: " Transfer of Technical Knowledge by International Corporations ", in *The American Economic Review* (Menasha, Wis.), Vol. LVI, No. 2, May 1966, pp. 259-67.

is a product-centred, semi-continuous flow technology. However, here the perspective is somewhat different: in place of indigenous entrepreneurs attempting to upgrade production methods, we have giant foreign firms trying to transfer a network of processes and procedures learned over a long period and within the context of a highly sophisticated technical and economic environment.

Baranson describes the specific problems encountered:

> On the whole, Cummins engineers found that machine labour was proficient at individual tasks—inefficiencies in manpower utilisation resulted largely from shortcomings in the over-all planning and co-ordination of production efforts. By Columbus standards, attentiveness to work at the Poona plant appeared lax, and there was an air of disorder with parts scattered about and clothing hanging near running machinery. Instead of working at a fairly steady pace, an Indian factory labourer works in spurts of energy and may then go off while a co-worker keeps an eye on his machine. There is a low level of literacy among machine labourers and some difficulty in handling blueprints, but once a routine is stabilised, the operation is handled with reasonable efficiency.[1]
>
> Recruitment of engineers and technicians is a problem of quality rather than quantity. The materials control laboratory and machine shop are, on the whole, adequately staffed. But there is a dearth of such personnel as foremen and shop managers with any depth of industrial experience. The deficiency is in the experience and know-how in setting up and co-ordinating segments of industrial operations. The labour shortages are in shop foremen, production control managers, and plant section managers with substantial industrial experience. Part of the deficiency is attributable to substantial differences in the Indian approach to industrial management.[2]

From Baranson's descriptions, one is led to conclude that the quality of work performance has played a significant part in the problems of the automotive industry. The cause does not appear to be primarily related to specific technical skills, but rather to co-ordination, quality control and input utilisation.

Ammonia-based fertilisers

This industry represents a large-scale continuous-process highly automated technology. Manufacture consists of the production of hydrogen with a steam-reforming process based on natural gas or naphtha, mixing with nitrogen, centrifugal compressing and synthesis into ammonia, reacting ammonia with a salt, and final processing. Important economies of scale (centrifugal compression, lower hydrocarbon feed rate and inert purge, storage economies, the 0.6 rule on capital costs) are realised up to outputs of 1,000 tons per day.

---

[1] Baranson, *Manufacturing Problems in India*, op. cit., p. 80.

[2] ibid., p. 81.

Raw material, fuel and capital charges dominate the cost of production: total labour and managerial costs are 4 per cent or less. Hence the industry's viability is not necessarily dependent upon the availability of local technical and organisational competencies—in the extreme case, a fertiliser plant could be operated entirely by expert foreign personnel without seriously raising costs. Economic viability is a function of scale, non-labour input prices, transportation savings vis-à-vis imports, and rate of capacity utilisation. Capacity utilisation in underdeveloped countries appears to be lower on average, mainly due to marketing and aggregate demand insufficiency, interrupted supplies of inputs and design defects in plant installation.

Rough skill profiles for the operation of a composite "average" ammonia-urea plant (about 400 tons per day) have been constructed by the United Nations Organisation for Industrial Development (UNIDO) for countries at three different levels of technological sophistication (see table 1).[1]

The increase in labourers in India, the Republic of Korea and Africa is lower than the increase in other categories. Engineers and chemists make up a higher proportion of total employment than in the Federal Republic of Germany, Japan or the United States.

The explanation is partly related to shift operation. In Africa it is recommended that the shift operating and maintenance foremen and their assistants, along with certain operators, should be graduate engineers or chemists. In India and the Republic of Korea only the first-class shift foremen need be graduates, whereas in the United States or Europe such men are usually skilled technicians.

Every aspect of fertiliser production—scheduling of inputs, production processes, materials handling, packaging—is fully automated. Work consists of instrument reading, valve setting and maintenance. Despite the limited and infrequent human intrusion upon the productive process, here is at least one instance where the skill requirements of an automated technology are extremely high. Fertiliser might be contrasted in this regard with cigarettes, beer and cement.

As indicated by the above skill profile, the majority of fertiliser factory employees must possess extensive scientific knowledge, including organic chemistry, physics, thermodynamics and the specific technology of fertiliser production. This widespread requirement of a comprehensive technical understanding stems from the high degree of integration between the various stages of production, the vulnerability of several of these stages to human

[1] UNIDO: *Estimation of Managerial and Technical Personnel Requirements in Selected Industries* (New York, United Nations, 1968), p. 58.

Table 1. Effect of level of technological sophistication on personnel requirements, ammonia-urea plant

Occupation	Germany (Fed. Rep.), Japan, United States	India, Rep. of Korea	Africa
Engineers [1] and chemists	40	79	152
Skilled technicians	168	212	372
Labourers	62	66	132
Total	270	357	656

[1] Mechanical, chemical, instrument and electrical, civil engineers.

misjudgement, the extreme costliness of going off-stream, and the safety hazards of high-pressure steam systems. In short, every participant must be able to recognise the full repercussions of his actions. In line with the earlier discussion in the first section of this paper, the skill requirements fall almost entirely into category 1, "specific technological skills". Only a few people at the top need be concerned with what we have described as internal production management functions.

MANPOWER TRAINING

Let us begin our discussion of training and education with specific technological skills. Perhaps the most common feature of manpower forecasting during the 1960s, and overpoweringly so in Africa, was to identify technician-level skills as the most binding bottleneck to more rapid development. This, in turn, has led to the expansion of government-run technical institutes, the operation of which has been severely hampered by staffing difficulties, diversion of the better students to university courses, and low graduation rates. Careful investigation reveals that in most occupational areas there is no relative shortage of true technician skills, as shown by wage-rate differentials in the private sector. The rate of return on each post-secondary school year spent in a university is several times that of a year spent in a technical institute: hence the lack of interest among potential technician students.

The great majority of intermediate level positions are filled by experienced craftsmen whose in-service training and known supervisory competence are readily accepted by employers as a substitute for the formal educational requirement.[1] A recent and very careful study on promotion patterns in

[1] Kilby, *Industrialisation in an Open Economy*, op. cit., p. 256.

16 Belgium and Argentinian firms (iron and steel, metalworking, electrical equipment, chemicals) showed that, under favourable conditions, 1.6 to 2.3 years of experience was equivalent to 1.0 year of full-time training for job qualification as mechanic, electrician, draughtsman, laboratory assistant and assistant engineer.[1] The implication of these facts is that the use of formal educational attainment to define skill level should be abandoned in manpower planning.

A related question to manpower planning is, who should operate the subprofessional technical education system. In Africa and a number of lesser developed Asian countries, where the system is largely publicly operated, a good case can be made for giving the private sector—specifically the large firms—a greater role. Not only would such a change result in the increased efficiency of technical training, but also the cost per student would be materially reduced and the government's administrative burden lightened.

Consider the case of craftsmen. By virtue of their closer contact with actual market scarcities, an employer-operated system would be more efficient in its selection of trades. Less encumbered by formal regulations and bureaucratic procedures, private firms can establish, expand, adapt or discontinue training courses with greater speed and at lower cost. Training orientation and content are automatically brought into close harmony with the requirements of the workplace and student motivation. Theory is effectively integrated with practice by greater proximity to the actual working situation and the immediate presence of a prospective employer. Moreover, after the first year, the trainee's contribution to output usually offsets his maintenance costs, which represent a significant share of the total training cost.

However, it is in the realm of instructor recruitment that the large firms possess their greatest advantage over alternative methods for organising technical training. By drawing instructors from their own career technical staff on a rotation basis (a long established and proven practice) the firms face neither a shortage of instructors nor the necessity to pay high differentials either because instructors are scarce or because they cannot be offered job security. This system of instructor recruitment has the additional merits of ensuring that training will be kept abreast of the most recent developments in applied technology and, by familiarising the senior technical staff with the special problems of comprehension and communication that arise in the learning process, of improving the latter's managerial effectiveness on return to regular duties.

---

[1] ILO: *Experience on the Job as a Substitute for Formal Training: An Empirical Study Based on Quantitative Analysis of Personnel Records in Sixteen Undertakings*, by J. Maton (Geneva, 1969; doc. D.7.1969; mimeographed). A condensed version of this document was printed in *International Labour Review*, Vol. 100, No. 3, Sep. 1969.

Administratively, shifting most of the trade training and the basic engineering technician courses to the employers (including government departments and public corporations) would involve government subsidies. Such subventions could be partially recovered, if deemed advisable, by imposing some form of per capita "apprentice tax" on all (or some) employers of skilled labour. In any event, the net savings to the public purse would be substantial.

About the formation of technological competencies at the university level and the development of market-oriented management skills (required for category 3), we have nothing to add to conventional wisdom.

The area of work performance of greatest concern in the general process of effectively absorbing more productive technologies is that of the activities connected with co-ordination, superintendence and control. Deficiencies in this area were illustrated by the farm implements and automotive industry studies. Progress here will have to come from many sources: greater emphasis on industrial engineering and production management in the curricula of technical institutes and schools of business administration, productivity demonstration extension service, subcontracting arrangements, fiscal incentives for high levels of utilisation of rated capacity, psychological training in the handling of authority relationships and the general inculcation of a philosophy of scientific management among industrial entrepreneurs.

The rationalisation of production has both a technical and a sociological aspect. Technically it is a question of diffusion of knowledge of industrial engineering procedures—the study of work methods, plant layout, materials flow, quality control procedures, and so on. Commitment to industrial efficiency and a sensitivity to opportunities for improving quality and removing slack must extend right down through the foreman level—hence the need for promulgating "Taylorism" at all levels from craft education upwards.

The human side of this rationalisation process involves the engineering of worker co-operation. This falls into two parts: the provision of financial incentives (e.g. payment by results) and constructive supervision. The first is easy to obtain and the second is very difficult. Training in supervisory techniques in underdeveloped countries is given in much the same way as in developed economies. The teaching methods are the case study and role-playing; decision making based on analysis, effective communication and warmth in human relations are stressed. The greater productivity from doing the job the new way (and the personal gain therefrom) is compared with that from the old non-analytical authoritarian approach. Back at the plant, after an interval of a few weeks, the authoritarian approach reigns as before.

A useful training method to develop the desired supervisory attitudes might be one that parallels David McClelland's need-for-achievement training. This would not only teach the techniques and benefits of modern supervision, but also enable the costs of this behaviour in terms of existing social norms to be faced squarely. By forgoing the right to remain aloof from the practical details of a subordinate's task, by treating social inferiors as equals, by sharing authority with a person of lower status through co-operative effort—by doing these things a supervisor is engaging in socially degrading activity which undermines his social worth, both in his own eyes and in those of his charges, to the extent that they both adhere to traditional norms. Hence the strong tendency to return to traditional modes of behaviour. By forcing the individual to see this conflict and to recognise the inconsistency between his professed managerial goal and his actual behaviour, we may hope to see, for at least a significant proportion, a rationalisation of supervisory behaviour consistent with modern status aspirations.

PRECONDITIONS FOR EFFECTIVE USE OF AUTOMATED TECHNOLOGY IN INDIA

ISHWAR DAYAL
Indian Institute of Management, Ahmedabad

The experience of Indian industry in using automated technology has been mixed; it has been successful in some companies and discouraging in others. In this paper we shall explore the conditions needed for the effective use of automated machinery in India through a brief survey of three such experiences.[1]

RESISTANCE TO CHANGE

A primary input in planning a strategy of change is the planner's assessment of the employees' likely response to the change. Recent consulting experience in a large public sector organisation in India supports the view that change is, by and large, accepted by employees under certain conditions. For instance, a new chief executive of one company with about 2,500 employees undertook a major programme of change. In the author's discussions with over 100 employees of this company, individually and in small groups, the most common feeling expressed at all levels was that change was essential for the organisation and need not be debated at any length. Their primary concern was *how* change would alter the present systems. Would employees become redundant? Would they be helped to learn new skills? Would their future promotion be affected by an intake of new people at senior levels? How would merit be determined?

The changes initiated in this organisation have created far more activity, a far greater desire for higher achievement, and much greater effort for learning about competitors than ever before. The trade union concerned has also recently signed an agreement whereby a number of occupational

[1] The author is grateful to his colleagues, Professors B. R. Sharma, Sudhir Kakar and J. G. Krishnayya, for comments on an earlier draft of this paper.

categories have now been abolished, and there is acceptance of mechanisation of certain jobs in the company.

In the airlines in India trade unions and employees alike have welcomed the Government's decision to buy jet and supersonic aircraft. The departments affected have themselves often suggested additional and improved uses for computers and other mechanical systems.

Neither these experiences nor our observation of behaviour suggest that in India resistance is a reflex response to change. A growing body of research evidence suggests that basic cultural factors, or those related to family background, do not depress the performance of employees in industrial roles. Employees, irrespective of their backgrounds, are able to adjust satisfactorily to their work roles in industrial occupations.[1] From the evidence of satisfactory adjustment to work roles one can hypothesise that technological change *per se* may be unacceptable because of factors within an organisation but not because it is human nature to resist change.

SURVEY OF EXPERIENCES

The three experiences we shall examine are intrinsically different. In the first case, ABC Corporation successfully employed modern technology. Within about five years of introducing change, however, the company was involved in prolonged strife mainly due to its failure to develop a viable management system.

The second case relates to the introduction of a computer in XYZ Bank. After the initial understanding between the management and the union, further utilisation of the computer has been blocked primarily because of social pressure from the reference group of employees.

In the third experience the management of KPE Corporation was able to integrate the human and technological requirements of an automatic system and, compared to the other two cases, has achieved stability.

ABC Corporation: introducing high-speed special-purpose machinery

The story of ABC Corporation is one of the successful use of modern technology introduced between 1954 and 1960, and of the failure to co-ordinate

[1] B. R. Sharma: "Commitment to Industrial Work: The Case of the Indian Automobile Worker", in *Indian Journal of Industrial Relations* (New Delhi), Vol. 4, No. 1, July 1968, pp. 3-32; N. R. Sheth: *The Social Framework of an Indian Factory* (Manchester University Press, 1968); I. Dayal and M. S. Saiyadain: *Predispositional Variables and Responses to Authority in an Industrialising Society* (Ahmedabad, Indian Institute of Management, 1969; mimeographed); R. D. Lambert: *Workers, Factories and Social Change in India* (Princeton University Press, 1963).

the subsystems needed to maintain the viability of the system as a whole. Various imbalances caused stresses and strains within the system and in relationships at all levels, leading to a six-month strike in 1963. We shall look briefly at the background of the firm and the problems that arose subsequent to its golden period.

ABC Corporation was set up in 1934 by an electrical engineer. He ran short of money within the first two years and invited a large industrial group to invest in the company. The new owners acquired 75 per cent of the shares in 1936 and the remaining shares two years later, and contributed top-level technical, finance and sales personnel from their subsidiaries.

During the war years ABC Corporation was engaged in manufacturing a variety of engineering products and showed satisfactory profits. Up to 1949 the company continued to manufacture fans, sewing machines, hurricane lanterns and press components. In 1949 the top management reviewed its operations and decided upon its future organisation and technical policy. The main elements of the company's policy and strategy since 1949 may be summarised as follows:

— to expand as rapidly as their resources and the market would allow;
— to "mass produce", reach economies of scale, and then sell at prices lower than those of competitors;
— to use modern industrial technology to achieve efficient production;
— to ensure productivity and labour peace through the employment and development of skilled personnel, rewarding them highly for production increases and recognising their rights to collective representation.

ABC Corporation decided to suspend the manufacture of all products except those that could be mass produced, and therefore concentrated on the manufacture of sewing machines and electric fans.

In 1949 the company reorganised its top management structure by appointing two general managers, one for the works and the other for finance and sales. Both general managers reported to the chairman, who was also head of the parent group.

The company's employee union had been registered in 1942, but the management had not recognised it [1] and the general manager had had no contact with the union leaders until 10 years after its registration. After a brief strike in 1952 the company recognised the union and started extensive consultations with its representatives on the company's plans, wages, service conditions, productivity, and issues that required co-operation by workers

[1] In India an employer may withhold recognition of the union even if it is registered under the Trade Unions Act.

Table 2. Sales, profits and dividends, ABC Corporation

Year ending March	Gross sales (millions of rupees)	Gross profits (millions of rupees)	Rate of equity dividend (%)
1955	22.39	1.99	12
1960	86.25	6.56	17
1961	108.47	8.78	17
1962	115.80	9.12	15
1963	134.44	7.44	10
1964	104.16	5.49	8

in the plant. Two years later the management and the union signed a five-year agreement which broadly covered the revision of wages, the production bonus and the establishment of joint consultative committees at the company and departmental levels. A number of joint committees for production, job evaluation and similar tasks were also set up.

Between 1955 and 1962 ABC Corporation made impressive progress and sustained steady growth in production and in domestic and export sales. Tables 2 and 3 give the company's financial results and employee earnings for selected years between 1955 and 1964.

Between 1954 and 1960 the general manager (works) initiated and carried out changes in the factory. These involved changes in both technology and organisation, including the following:

— introducing high-speed special-purpose machinery for the manufacture of components;

— modernising the machinery, equipment and methods in the machine shop, foundry, paint shop and other production departments;

— establishing new departments such as industrial engineering, planning, design and development, and so on;

— consulting with the union and joint decision making in matters of rates for incentive payment for new tasks, employee amenities and such matters as needed the employees' co-operation;

— establishing paternalistic relationships with employees and helping those who came to the general manager with personal difficulties.

The company made a unique departure from traditional labour sources, employing literate workers and training them for semi-skilled and skilled jobs in the plant. The junior supervisory positions were filled mainly from within the organisation by promotion. The specialist and some middle- and senior-level managerial positions were filled by recruitment from the

Table 3. Employment and earnings and output per employee, ABC Corporation

Year ending March	Total number of employees	Annual earnings per employee (rupees)	Index of real earnings per employee	Index of physical output per employee
1960	4 670	5 102	100	100
1961	4 967	6 021	120	133
1962	5 070	5 793	108	123
1963	5 367	6 296	113	141

open market. The company did not experience any difficulty in finding suitable people for either technical or non-technical jobs.

During the late 1950s the efficiency per employee was high. It compared favourably with similar plants in Europe. The earnings, based on a production incentive scheme, went up roughly three times, and far exceeded those of comparable jobs in other industries in the area.

In 1963 ABC Corporation faced a prolonged strike over a charter of demands by employees for further increases in wages and the annual bonus.[1] Among the reasons for the strike were the following:

1. The financial position of the company had been deteriorating progressively: high inventories of components, raw materials, semi-finished goods, increasingly higher payments to workers for higher output defeated by poor production scheduling that locked up capital . . . all these took their toll, and the liquidity position of the company reached a critical level in 1961 and 1962.

2. There were consultative systems in the plant, but decisions were taken only by the general manager. Within two years of setting up the consultative machinery, all lower-level committees became useless. Due to the attitude of the general manager, the union made a habit of approaching him directly. All levels of managers and supervisors became frustrated and ineffective in their jobs. Moreover, the general manager often changed the decisions of managers when the union approached him. The major elements of the general manager's methods included:

— centralised decision making;

— quick ad hoc decisions, often verbal and unrecorded;

[1] Bonus is declared by a company according to its operating results. Based on a formula, a certain percentage of the retained profits is payable to employees. In the case of ABC Corporation the difference between the management and the employees was over a demand of Rs 30 (= US$ 4) per employee.

— lack of systematic use of the specialists in the organisation;

— development of a network of personal relationships between the chairman, the general manager and the workers;

— dealing with departments individually, and conspicuous absence of a management team; poor co-ordination between departments.

The general manager showed that his first concern was for production. He frequently exhorted workers to produce more and rewarded them by giving a high production bonus. The prevailing impression among employees at all levels was that the central objective of ABC Corporation's activity was to increase production and maintain it at a high level. Other aims were subsidiary, and other problems would be resolved, provided output was high. This predominant factor formed the basis for dealing with recurring disciplinary, technical and other issues in the departments.

3. A standard way of pressurising management to agree to workers' demands was to stop production. The managers were taken to task by the general manager if the output in their departments was low. The workers used this as a lever for making departments grant favours.

These were the significant characteristics of the situation in ABC Corporation. Many other factors also contributed to the failure of the system: power struggles at the top levels of management; failure of interpersonal relationships among the senior management staff; almost complete lack of institutionalised practices in administration; alienation of employees at managerial and worker levels, and so on.

The company had originally achieved a substantial measure of success in introducing complex production technology without being confronted by resistance. It had been able to acquire the capacity to develop new products and to compete aggressively in the home and international markets. The employees accepted the changes and were generally proud to work in a "modern plant". They spoke well of "their" organisation in public during the period of change and innovation. Interviews and the minutes of the meetings between the management and the union show that employees gladly accepted the modern machinery, showed interest in learning new skills and attained a level of production far beyond the expectations of the management. But the achievement could not be sustained by the system as the serious breakdown of 1963 showed.

We have discussed this situation in order to illustrate two points: (*a*) the introduction of new technology invited neither resistance to change nor diffidence in adjusting to new machinery and equipment; (*b*) the failure to change administrative practices (such as planning, control, supervision, etc.)

inhibited full utilisation of modern technology. The imbalances within the organisation caused by inappropriate management systems led to continued conflict and long periods of work stoppage. The findings of the research team suggested that the intensity of conflict in ABC Corporation would have been minimised if new technology had not been introduced or if, after introducing it, corresponding changes in the management system had been brought about.

In the last section of this paper, I have used the experience of ABC Corporation to emphasise that effective utilisation of automated technology includes the recognition and development of appropriate management systems.

XYZ Bank: union response to a computer

With approximately 1,600 branches spread throughout the country, XYZ Bank employs about 54,000 employees and is the largest commercial bank in India.

After coming to an initial understanding over the installation of a computer, the management and the union are now finding it difficult to agree on extending its scope. This is due mainly to the union's complex network of intra- and inter-organisational relationships. This experience highlights the external influences exerting pressure on the union to resist the extension of the computer's use.

By 1965 bank officers had concluded that the job of reconciling interbranch accounts could not be handled manually, and the management decided to install an IBM 1401-1410 computer.

Before ordering the computer, the management held prolonged discussions with the union and were able to persuade its leaders that, unless the interbranch accounts were computerised, the bank would not be able to open many new branches. Before deciding to buy a computer, the bank had tried IBM data processing equipment in four centres without success. Hence the management's arguments were supported by the experience acquired from trying alternative ways of coping with the work and by convincing evidence that the bank's present and intended size required a computer.

Unions in the banking industry. In order to understand the union's dilemma, some knowledge of the background of unions in the banking industry is necessary. Banking has the largest white-collar organised unions in the country. The strength of the unions has grown in the past 30 years. Between the late 1940s and the early 1960s salary and service condition awards had been made by tribunals appointed by the Government. These awards applied to the banking industry as a whole.[1]

In the early 1960s the managements of non-nationalised banks and the national unions negotiated a bipartite settlement for the first time. Bipartite negotiations were held at all-India level between the Indian Banks Association (representing the employers) and the three national unions of employees.[2]

In the banking industry three levels of union organisation have developed. First, there are the three national unions of bank employees. Second, units organised at different branches of the same bank are joined together in an all-India federation, affiliated to one of the three national unions. Third, regional organisations have been formed, consisting of units at a bank's branches within a particular geographical area. In addition to these three tiers there exists in each of the states an additional union organisation, with an inter-union membership.[3]

The picture is somewhat more complex than stated here. There is more than one union within each bank, each being affiliated independently either to one of the national organisations or to none. There is thus a complex network of relationships within the unions. Often the management of a bank has to deal with more than one union.

The key issues of salary and service conditions are negotiated at the national level of the employers and the employees collectively. The bargaining strength of each union depends upon its existing membership and its capacity to attract more members into its fold. Each union has to compete with other unions for membership from the same population of bank employees. Each federation negotiates separately with the representatives of the bank management. The negotiations are held concurrently with each of the federations; and no agreement is reached unless all three agree to the final proposals.

There is keen competition among bank unions to retain their membership both at the national level and among the affiliated unions. This type of competition imposes upon the unions the need for cautious policies in so far as they have to depend upon popular appeal to obtain membership.

---

[1] The Reserve Bank of India and the State Bank of India, which were national banks, negotiated their own agreements with their unions and were not a party to those mentioned above. In 1969-70, 14 other banks were nationalised by the Government.

[2] All-India Bank Employees Association, National Organisation of Bank Workers, All-India Bank Employees Federation. The State Bank of India is a separate body with two tiers: all-India and regional organisations.

[3] These organisations are not generally concerned with issues relating to wages or service conditions, but stimulate other union units within the state to support the efforts of a particular union when engaged in a dispute.

Two statements representative of the views of leaders of bank unions are:

> We have to support even weak cases of indiscipline in the bank; if we didn't, we would lose a member while some other union would still take up the cases for discussion with the management.
>
> Now that the other union is being organised in the bank I have to handle every issue very carefully. For the last three years I have had to spend time fighting a battle on two fronts—the management and the rival union.

Each unit-level union also has to maintain a fine balance between the directives of its national federation, the all-India federation of its own bank employees, and regional membership. The decision making is often a function of balancing the divergent cues from different publics. The necessity to do this introduces some rigidities into the system, which thus tends to restrict independence and freedom of deviating from policies that might be less popular than those of the competing union.

Two representative statements are:

> One national union has taken a decision to oppose automation. Some of our members are not opposed to automation but we have to make a solid front against the management; they would otherwise take advantage of our differences.
>
> We have special difficulty in negotiating with our unions because we have four unions and what is agreeable to one is opposed by the others. (A chief bank executive)

The national federations have taken a policy decision to oppose the introduction of computers. The reasoning behind this decision was that computers would displace men working in banks and reduce the employment potential of the educated unemployed in India. They argued that a poor country should use manpower instead of expensive equipment.

Introducing a computer. Since XYZ Bank was one of the largest banks and because of the size of its operational problems, the union was willing to give qualified acceptance to the installation of the computer. At the same time it stipulated certain specific uses for it, for instance, the reconciliation of interbranch accounts and the preparation of such statistical information as the management might want from time to time, provided that this work was not already being done by one of the existing departments.

As the computer began to function in XYZ Bank, the union was subjected to strong adverse pressures from the unions in other banks. The leadership was criticised for having agreed to a proposition that other unions in the banking industry had stoutly refused to entertain. Although an explicit policy decision by these other unions to oppose computerisation was made only after the agreement in XYZ Bank, the union leadership was nevertheless criticised for the agreement. In some parts of the country a splinter group in the union used this decision to plan a competing union.

Thus the pressures upon XYZ Bank's union were from both their own splinter group and other unions at the national and regional levels. The union leader could resist the pressures because of his 20 years of leadership and the services he had rendered to his union throughout this period. Nevertheless, the computer was a sensitive issue, and the union had to take this criticism into account when negotiating with management over other uses for it.

After gaining experience in using the computer over about three years, the management felt that they should extend its use. Both management and union were reluctant to precipitate the issue and have allowed it to drag on for many months, waiting until the time seems ripe for a full-scale discussion.

The following characteristics of XYZ Bank's experience are of special interest. In the first place, the initiative for introducing a computer came from the management. The compelling circumstances of the situation had to be recognised by the union, and its leaders had to agree to the introduction of the computer. Its use, however, was defined as that aspect of bank operations which experience showed could not be manually performed.

The union had to find a balance between the internal reality of the bank's operations and the external social reality (their reference group) in their acceptance of a computer. The use of the computer and the extension of its use depend largely upon the balance between these two factors. Note that specific manpower questions such as lay-off procedures, hiring policies, wage levels and retraining procedures, as related to the new technology, were not a source of friction and therefore not important in shaping the attitudes of union members and their leaders. In the Indian setting companies can easily afford to be generous on all those points. XYZ Bank would have to be more than generous, however, granting truly uncommon advantages if the union was to withstand the hostility of the reference group. Change in the reference group attitudes would, of course, be another way of getting wider use of computers.

KPE Corporation: developing an organisation

KP Engineering Corporation (KPE) is a new organisation set up in the mid-1960s in Bombay to manufacture electrodes. The writer served in the company as a consultant for about 15 months during the earlier stages of its development. Its relevance to the present paper is that high-speed automatic equipment was installed in a plant which had to compete on quality, price and profitability with already well-established companies.

KPE is an example of collaboration between a European and an Indian business house. The Board of Directors had equal representation, although the present chairman and the executive director are from among the Indian collaborators. The collaboration agreement lays down the specific tasks for the Board: important operating decisions on financial matters, sales and manufacturing policies, recruitment of key personnel and review of the company's progress. The Board members did not interfere with the day-to-day working of the company and left the management of KPE to its chief executive.

From the beginning, the management of this company attempted to develop managerial systems and employee attitudes for exploiting technological advantages over competitors.

Due to the highly automated production system the total strength of workers per shift was small, but a number of workers had to handle more than one operation. The efficiency of the plant depended upon the awareness of these workers of the advantages of speed, a low percentage of rejects and the fulfilment of prescribed routine maintenance schedules in order to avoid idling of the machines.

Two other important characteristics of the production system were: (*a*) raw materials were imported and, as they represented about 60 per cent of the cost of production, wastage and rejects had to be as low as possible; (*b*) indigenous sources of supply had to be found to substitute for imported components as soon as possible. The workers had to work with extra care to try experimental and sometimes repeated runs. In short, the general manager wanted KPE employees to show a high regard for machine technology, for speed and for quality of product, and wanted the management to generate conditions in which employees could be involved in their work.

The responsibility for normal production had to rest with operatives rather than with the supervisors, who could provide only general, as against close, supervision. This type of demand upon workers was somewhat different from that of most production systems from which the employees were drawn.

In the initial stages the top management team broadly analysed the problems of the company and defined the type of organisation they wanted to develop. The following were a few representative ambitions of the group:

— people working in the company should be happy and contented and should find maximum scope for their abilities;

— conditions should be such that people work willingly without need for coercion;

— KPE should gain a high public reputation and employees should feel proud to belong to the company;

— KPE should be a place where people co-operate with one another;

— KPE should have the capacity to use information and make data available for effective decision making.

The main features of the development programme at KPE are the following:

1. Six months before the target date for starting production, the company appointed 25 per cent more trainees than the estimated number of workers required. The trainees were placed in another engineering factory of the Indian collaborators for training in the maintenance and production departments. The training was supervised by the production manager and the personnel officer of KPE, who maintained close liaison with the employees. At the time when production at KPE was begun, a final selection of workers was made from the trainees. Provision for this had been made in the letter of appointment at the time of recruitment.

2. With the help of the consultants, the top management group of eight worked intensively through the problems of KPE at individual, group and organisational levels. Examples of some of the main issues are:

— at the individual level, members' concerns were for their identity in the company, for job security, and as to whether they would be isolated from important activity (alienation);

— at the group level, their concern was to know how their tasks would fit in with those of others in the management group (interdependence). What kinds of information did each require to do his job and what information should he provide and with what frequency? What were the obligations to each other in their roles? Among the owners, representatives of both Indian and European collaborators, there was concern about establishing relations and about the type of superior/subordinate relations acceptable in KPE compared with those they had experienced elsewhere;

— at the organisational level, there was much concern about the development of over-all policies for the company. What kind of organisation would be best for KPE? How should subordinates be trained to take up responsibility? What kinds of control systems, methods for evaluation of employees, review procedures, decision rules should be instituted? How should demands deriving from the external environment, from government regulations and their implementation agencies, from other employers, from competition, and so on, be met?

With the help of the consultant, the top management systematically worked on behavioural and organisational aspects of their work. They clarified role interdependence, mutual expectations and obligations towards one another. They developed such administrative details as would help them to perform their duties satisfactorily. Each manager, in turn, began to examine similar issues with his own subordinate and so on down the line.

During the first 15 months the top management carefully developed the steps needed to attain the goals they had set for KPE in the beginning. Though this is not a change programme in the sense that ABC Corporation was, the experience of KPE is significant in so far as the strategy of the management aimed at developing, within the organisation, stable systems that would sustain growth as well. Through conscious effort to evaluate the requirements of the new technology, and by developing corresponding managerial systems for its use, the management was able to start production earlier than anticipated and to experiment with cheaper raw materials.

Some representative comments of the collaborators and the top management in the context of the organisational development programmes will indicate the results:

- of all the welding electrode factories set up by the European collaborators in other countries, KPE has achieved the required level of production and quality in the shortest time;
- KPE started work with top management personnel who had limited prior work experience;
- there have been serious conflicts in KPE but the top team now works well;
- some confusion remains in daily work because of delays in setting up proper systems and procedures in KPE ("Quite frequently we get into an awful mess; but we are confident now that things can be sorted out soon");
- much remains to be done to build up each department vertically and to develop co-operation at lower levels.

SOME GENERALISATIONS

From the experiences we have discussed, we may make certain generalisations relevant to the use of automatic technology in Indian industry. Technological change is likely to be of advantage under the following conditions:

1. The strategy of change attempts to reduce, as far as possible, the incongruence between individual and organisational goals. ABC Corporation achieved this by enabling the workers to earn more in their jobs, by involving

them in the process of change, and by defining corporate goals perceived by employees as being consistent with social objectives.

2. The managerial systems are explicitly developed so as to be consistent with the demands made by automatic technology. This aspect includes both structure and the administrative processes of the firm. The experience of KPE Corporation suggests that the management attempted to develop an explicit system congruent with the demands of technology and the goals they wished to achieve.

3. Employee acceptance of new technology is observed to be an important condition for its success, and is so perceived by the reference group of the concerned union. The influence of the reference group and the social pressures to which the union is subjected could minimise the advantages of automatic technology if management were not sensitive to the reality of the social links between the union and its national affiliates. XYZ Bank had to restrict its use of the computer because of the competing pressures on its union by other bank unions.

MANPOWER AND EFFECTIVE UTILISATION OF AUTOMATIC TECHNOLOGY IN INDIA

In India it is not common for automated machinery to replace an existing plant or equipment. Isolated pieces of new machinery are often installed in the old plants. A few people needed to run such equipment are selected from within the organisation, or occasionally recruited from the open market.

The new plants are predominantly process plants. For these, educated people (of whom there are many in the country) can be trained. This is also true as regards maintenance and other technical jobs.

Most computers in use for commercial applications carry out routine analysis, making limited demands upon designing and programming skills.

Systems analysts and programmers for the existing needs of computer installations are available. For more sophisticated applications, familiarity with the equipment is naturally needed. As second- and third-generation computers become available in India, more expertise will be needed, and a longer lead time for training may be required for developmental and programming work.[1] Such expertise is at present available mainly in research and teaching institutions and in a few of the larger business organisations.

[1] The first electronic digital computer for business application was installed in 1961 although the first computer was installed in India for a research organisation in 1956. The Government of India has issued licences to four manufacturers to set up manufacturing facilities in India. Within the next few years computers in the 1401 series and the 1901 (third-generation) series will become available.

Manpower requirements

The manpower requirements of an automated plant are at the levels of: (*a*) operatives or technicians; (*b*) specialists in the technology introduced; (*c*) supervisory and managerial staff with an appreciation of the technology, its applications and the newer ways of managing related activities and people.

In India, almost without exception, the operatives and technicians employed on automated machinery have been trained by organisations from within. As the installations of such machinery (including numerical-control machine tools) have been few, no serious shortages of trainable operatives are reported. The more important skills in such instances are for maintenance, not operating. In the past 20 years the technical schools and colleges have made available to industry sufficiently large numbers of engineering or technical skills, and no serious shortages occur. The rate of growth in automated machine tools or process machinery is likely to be small, and the writer does not visualise manpower shortages for such limited expansion. For future growth, in 10 years or so, the nucleus of existing installations and training institutions would yield adequate numbers of skilled personnel.

The experience of introducing electronic data processing equipment in India also suggests that most companies have obtained the necessary manpower from within their own organisations and trained such personnel initially with the help of the suppliers, and in later phases, from within. In a survey of about 100 companies who use computers (carried out by the Indian Institute of Management, Ahmedabad), 35 companies were taken up for detailed study. It was found that an overwhelming number of the companies used computers for routine data processing purposes. Two companies were using electronic data processing for non-routine analysis.[1] For programming routine commercial applications, most companies have been able to obtain programmers from within their organisations or, in a few cases, from the open market.

The evidence suggests that manpower shortages are due primarily to lack of planning rather than to non-availability of trainable staff. As third- and fourth-generation computers come into use and more sophisticated applications are made, more planning for training will be necessary, the basic skills being already available in the country.

The educational and research institutions have fairly advanced computer skills. The writer believes that adequate specialist services are available

[1] G. R. Kulkarni: "Computers—Experience of Indian Companies", in *IIM Alumni Conference Papers* (Ahmedabad, 1970).

for sophisticated applications of electronic data processing. Within business organisations, increasing numbers of employees are also becoming available who can deal with the complex uses of computers.

The non-availability of manpower at operative, technician, and specialist level is unlikely to be the most important handicap to Indian business in using computers. The shortage in supervisory and managerial personnel is, however, acute.

Effective use of automatic technology requires change in organisational structures and administrative practices. Centralised and paternalistic patterns of management prevent potentially good supervisors and managers from developing.

Attitudes of employees

One must distinguish attitudes that relate to automatic technology from those that are generally needed for any industrial work. An interesting example from a company in Bombay might help to indicate what special attitudes are needed after automation.

A light engineering company, which has been a leader in the packaging business for many years, decided to set up a second plant in Bombay. The existing machinery in the plant consisted mainly of manually and mechanically operated machine tools, with a few automatic ones. The additions to older machinery were made whenever import licences were obtained and technically more sophisticated machinery could be bought abroad.

In 1964 the company was able to secure import licences for automatic machinery. In deciding to construct a second factory in Bombay, the general manager had a choice between duplicating the first plant's operations in the second factory and equipping the second plant with only automatic machinery and processes. He decided to adopt the second course.

The strongest point in favour of this decision lay in his observation that the attitude of employees working on high-speed or automatic machines is different from that of employees working on older types of machinery. The utilisation of automatic machinery, he said, depends primarily on how conscious the employees are of *speed* and *time*. This consciousness is needed among employees on the factory floor as well as in the supporting services in the plant. A mixture of manually operated and automatic machinery would certainly tend to develop a mix of employee responses unsuited for the full use of automatic machinery.

Another chief executive, referring to the development of new attitudes, said: "We have to get used to the idea of sending quotations by telex if the inquiry has come by telex. If the customer had wanted to wait he would have used either the slower postal service or a telegram. This is the substance of what I want my people to understand about the change we have undertaken to bring about in this organisation."

SUMMARY

This paper has been concerned with locating conditions for the effective use of automated technology in India. It claims that the effective use of automation depends upon the strategy of introducing change, on its acceptance by employees, and on internal and external factors that influence the organisation.

The strategies of change which assume that employees resist change may be based on incomplete evidence about behaviour. Change is endemic to modern society, and most people consciously or inadvertently adjust to change. Employees resist change that goes against religious and ethical values, not change that is primarily work-oriented. It is first necessary to determine the conditions in which employees resist change, and then a suitable change strategy can lead employees to accept new technology.

Three experiences of Indian companies in India have been discussed. In one case the use of new technology was at first highly successful; but within five years the company faced a major strike and has not yet recovered from its aftermath. The most important reason for this failure was the management's neglecting to develop an appropriate organisational structure and administrative practices consistent with the requirements of the technology.

In the second case the bank introduced a computer for the reconciliation of interbranch accounts and for statistics. Due to social pressures to which the union was subjected, the bank could not extend the use of the computer.

In the third case the management consciously tried to develop a managing system by intensive exploration of social and organisational processes and to achieve a synthesis between the requirements of the technology and the organisational goals.

In the Indian context the effective use of automatic technology depends upon the managements' concern with both social and economic factors and not with the economic factor alone.

The effective use of technology depends on the introduction of a programme for change which should have the following essential characteristics:

— a demonstrable effort prior to the change to improve the organisational structure, administrative practices, and intergroup and interpersonal relationships;

- besides a careful analysis of the economic rationale of automation, an effort to enlist support for this rationale with the trade union at the unit level and, where necessary, with the reference group trade union organisation;
- planning of change with specific reference to: (*a*) establishing goals for employees and management in introducing new technology; (*b*) employee gains from the new technology; (*c*) training and retraining of the employees at operative, supervisory and managerial levels.

UNIONISM, TECHNOLOGICAL CHANGE AND AUTOMATION IN LESS DEVELOPED COUNTRIES

EVERETT KASSALOW
Wisconsin University, Madison

By virtue of the very problems it considers, this paper is tentative, speculative and in the nature of a survey rather than definitive. In the first place, any effort to deal with all less developed countries must be general and relatively superficial. Second, any effort to isolate the influence of automation and technology upon the formation and functioning of trade unionism is, to a significant degree, arbitrary and abstract.

It is one of the premises of this paper that it is possible to identify a substantial body of common social, economic and political characteristics and problems in a large number of Asian and African societies, despite their great geographical, social and demographic differences. These common characteristics permit some general conclusions to be drawn about the impact of technology and automation on the forms and policies of unions in these countries.

To begin, we shall consider the broad differences in patterns of worker response to technological change as they may be related to particular areas. Next, the paper will treat some effects of technology and economic organisation upon union forms. This is followed by some brief comments on labour relations at the enterprise level as related to advanced technology. Finally, some special consideration is given to the particular problems of white-collar workers and of unions confronted with automation in less developed countries. If we concentrate here on generalisations, we trust the reader will keep in mind the limits of such generalisations; to cite exceptions constantly would make the text hard to read and excessive in length.

We shall, for the most part, be treating automation along with other forms of major technological change, since there has been relatively little "pure" automation in the less developed countries. In some of these countries fine distinctions between rationalisation and automation are not always made.[1] A few specific "automation" cases will, however, be singled out for comment.

Finally, the paper is also tentative and speculative because the degree of research which a truly "scientific" study of such a subject would call for was simply not possible within the limits of the time and resources available. Some parts are based on the writer's own observations, others on long reading of the labour and industrial relations literature of less developed countries over a number of years, although only a few specific citations have been included. In addition, the writer has recently had the opportunity to discuss the subject with a number of scholars, government officials and trade unionists with long experience in the less developed countries; these conversations (and letters) have been most helpful, and indeed he could not have attempted such a "far-flung", if tentative, paper without the benefit of the thoughts of so many friends.

AREA DIFFERENCES IN WORKER RESPONSE

It is interesting to observe that the introduction of technological change including automated processes has provoked almost no major protest or concern on the part of African unions. This contrasts sharply with experience in such an underdeveloped Asian country as India [2], and, to some extent, Japan. While the latter country has emerged among the developed nations in the past decade or two, its industrial relations and union history can still shed some useful light on the experience of less developed countries.

Substantial Indian union resistance to technological change has occurred on a number of occasions. For example, resistance to rationalisation in the textile industry in the 1950s led to the conclusion of an elaborate special agreement to "control" the introduction of new machinery and equipment. In recent years proposals to computerise the Government's large-scale insurance company operations have provoked a bitter industrial relations conflict,

[1] "There is no precise definition for automation. The term is loosely applied in India nowadays to include any form of rationalisation of work process, mechanisation of work or adoption of improved methods which reduces the number of persons presently engaged in performing a specified work. According to this view, automation is an attitude, a philosophy of production, rather than a particular technology of electronic devices, and it is distinguished by only one feature, namely reduction in the number of persons who are to be employed producing a certain quantity of goods or services." Indian Ministry of Labour and Employment: *Automation in India*, paper prepared for the Indian Standing Labour Conference, 28th Session, New Delhi, 18 July 1968, p. 1.

[2] We have used a number of examples from Indian experience in this paper. India is one of the best documented of the less developed countries, and any student of labour conditions must tap this experience. Moreover, although it remains less developed, India has a large industrial sector and a relatively long industrial history.

including some "absolute" union opposition to all automation. Other cases can be cited of resistance to technological change, resistance which goes beyond what the student of industrial relations is familiar with even in the case of deeply entrenched older unions in developed countries. Resistance to advanced technology and automation often takes on an ideological tone among a variety of unions in India—left, right and centre—although, admittedly, all-out anti-rationalisation and automation campaigns are more often led by left-wing unions.

Almost by definition, a Marxist-oriented union movement in a non-communist State is likely to view with suspicion movements toward rationalisation and automation. The communist-oriented All-India Trade Union Congress has, for example, declared:

> The trade unions are not opposed to technological improvements but in today's context technological innovations become nothing but an instrument in the hands of big employers with large capital to spare, to enhance their already higher rate of profits without conferring any gain on consumers or society in general, either in the matter of prices or supply, and hence they become objectively an anti-social act in the path of man's progress towards freedom from poverty and drudgery of work. It only results in reduction of employment and wages. . . . Government should ban import and use of electronic computers and automation equipment for offices and factories.[1]

Indian reaction to technology is probably enhanced by the egalitarian atmosphere which has surrounded India's modernisation plans since Independence. The earlier Gandhian stress on the values of cottage-type industry may also influence the technological atmosphere in the country.

While resistance to rationalisation has not been a severe problem in Japan, the frequent *general* attacks on industrial modernisation coming from the country's largest labour federation (the SOHYO, also a Marxist-oriented union) sound somewhat like those heard in India. At the enterprise level, on the other hand, a high degree of co-operation characterises Japanese industrial relations (some would stress the paternalistic flavour of these enterprise relationships), and little or no serious resistance to technological change or automation seems to occur, with a few notable exceptions such as the coal mines. The need to keep products and prices competitive in foreign markets seems to have been a widely accepted pressure in favour of rationalisation in Japan. The language of labour protest at the national confederal union level in Japan may sound similar to much of what is heard in India, but concrete union reactions to technological change tend to be different.[2]

---

[1] *Trade Union Record* (New Delhi), 20 June 1966.

[2] In a 1965 statement the SOHYO spoke of its "fight against capitalist rationalisation". However, the tactics to be employed were more or less "conventional" in industrial relations parlance, namely "the demand for reduction of working hours" to offset any possible

This experience in Asia, and especially India, contrasts with that in Africa. Here, as we have said, one can find virtually no serious discussion of the possible demerits of technological change, automation, and so on.

How can we account for these differences?

In the first place, it is probable that the heavy population pressures and relatively unfavourable land/man ratios in a nation such as India, as compared to virtually the whole of Africa, make the problem of jobs a difficult issue, as well as the possible disappearance of jobs in a given plant or industry (even where counterbalanced by growth elsewhere). By tradition, some of the older fears arising from population pressure on land supply may still carry over in Japan since it is only a few decades since almost half of the Japanese population was employed on the land. Contrast this with Malaya, where land/man ratios have been more favourable and where little has been heard from the trade unions about rationalisation. One hastens to add that other factors also differentiate Malayan development, but nevertheless there is obviously no single Asian pattern.

It would also appear that the very "age" of industrialisation in India (and Japan), as contrasted with Africa, may help to account for some of the worker-union reaction to technological change. In both of these countries industrialisation goes well back into the nineteenth century, while in Africa it has barely begun. Africa has not nearly so long a memory of the early miseries of the factory system and urban life and so on.[1]

Along with the "older age" of Indian (and Japanese) industrialism has gone its identification with a traditional native bourgeois-proletarian class structure, which in most industrialising countries, in the earlier stages of development, has given rise to a deep sense of conflict over industrial relations issues.

The absence of a significant native industrial bourgeoisie in most of Africa (where the bulk of important industry has been owned by foreign capital) at least shifts the focus of possible working-class reaction. Here, historically, industrial relations conflict has tended to be caught up with anti-foreign agitation. Now, as most new African governments themselves

---

employment losses and keep a "favourable . . . situation of the labour market" (*SOHYO News*, 25 Jan. 1966). More generally, the SOHYO's objectives vis-à-vis rationalisation, while couched in radical phrases, follow the fairly standard practice of unions in many developed industrial societies operating within a market framework. In a 1963 handbook of *Principles and Activities* (p. 37), the SOHYO stated (in connection with modernisation) that its "major demand" was "(*a*) to reduce hours of work; (*b*) not to discharge workers; (*c*) not to worsen working conditions when workers are transferred to other places of work".

[1] Worker-enterprise relationships in Africa bear other scars, however, notably those stemming from the racial gap between workers and management in colonial Africa, the earlier patterns of forced labour, head taxes, etc.

assume the task of leading and planning the industrialisation, consequent problems may not resemble the pattern of class hostility common elsewhere.

In contrast to earlier industrial experience, it is almost as though the present-day industrialisation process is "de-ideologised" in Africa. It is likely, however, that these new structures of industrialisation will produce their own tension and strains as development proceeds. It is interesting to observe that in the first years of their independence many new African nations find more industrial relations problems and strikes among public employees, including teachers, than among workers in modern industries such as mining and factories.

Development is relatively so recent in Africa that there is not the same historical experience on the part of the emerging working class which might cause resistance or fear of technological change in a manner similar to most earlier industrialisers. New plants and industries in Africa (except for transport and mines) are almost totally new, and there is little or no fear that older status or jobs are being wiped out, or that particular individuals are being displaced. Each new plant seems to be totally additive to the number of job opportunities. Again, there is often no worker experience, save with modern technologies—therefore there is no reason to resist them.

The long history of industrialisation in India provides a different social background. The textile industry, for example, had its beginnings in the first half of the nineteenth century and has been a present-day source of industrial relations friction as regards rationalisation programmes. It was an important source of a distinct Indian bourgeois and working class in several major cities. This same industry produced the first enduring stream of Indian unionism (beginning in Ahmedabad) over 50 years ago.

Indian textile workers have had experience with older technologies and can pose alternatives to new ones, particularly if they see any possible loss of jobs. As we have already noted, a major movement for rationalisation in the Indian textile industry (including the substitution of automatic machinery) provoked considerable union opposition in the early 1950s. A tripartite working party produced an elaborate agreement designed to guard against excessive unemployment effects and "to reconcile the conflict and facilitate the progress of rationalisation [with] . . . safeguards".[1] It provided for: joint standardisation of work loads; a halting of fresh recruitment when rationalisation was contemplated, with no filling of vacancies due to death and retirement (so as to provide slots for displaced workers); that surplus workers should be offered jobs in other departments wherever possible;

[1] This account is based upon Charles A. Myers: "Labour Problems of Rationalisation: The Experience of India", in *International Labour Review*, Vol. LXXIII, No. 5, May 1956, pp. 431-50.

severance pay for workers voluntarily retiring; seniority to prevail in retention of force; retraining in alternative occupations for workers laid off as a result of rationalisation, including support during training for up to nine months, with maintenance of the workers at management expense, and training at the Government's expense; full use of surplus labour in various projects undertaken by the Government; and sharing by workers in the gains from rationalisation, the object being to facilitate the retained workers' attaining a living wage standard through acceptance of rationalisation.[1] The unions were to be notified of proposed changes in technology and to be consulted before its introduction. Despite this general agreement, rationalisation did not make much progress in the Indian textile industry, as worker resistance at the plant level proved to be considerable.

Later, a new governmental committee advocated a phased introduction of new automatic looms over a period of 20 years, with an estimated displacement of 80,000 weavers. The union (TLA) opposed the plan and rationalisation made only slow progress, despite concern on the part of both Government and industry about Japanese and other foreign competition.

The existence of plural unionism in India, with frequently more than one union representing the workers in the same plant or industry, illustrates another of the difficulties besetting many less developed countries as they cope with industrial relations problems. Even if one union may be disposed to come to terms with rationalisation, a rival union is likely to take a more militant position of opposition. The result may be that the more "reasonable" union is also driven to oppose the rationalisation or automation effort.[2]

The situation in Africa, where plural unionism in a given shop or industry tends today to be the exception rather than the rule (as in the case of India), may also help explain the absence of friction over technological change.

But to refer again to the problem of the "age" of the industrial process, in the case of India it is of interest that very little resistance to rationalisation and the like has been shown in such new industries as machine tools, the engineering industry (including electronics) as a whole, as well as air transport. In these newer industries the workers appear accustomed from the start to higher and more dynamic technology. They operate, moreover, in an atmosphere of expanding job opportunities, and tension over rationalisation appears to be largely absent. For example, the successive introduction of

[1] This agreement was embodied as paragraph 60 in India's First Five-Year Plan, 1952.

[2] During the textile rationalisation difficulties of the 1950s one employer is said to have remarked: "We think we can get the INTUC leaders to adopt a joint approach—they do in principle. But their position is complicated by communist and even socialist opposition" (Myers, op. cit., p. 441). On other occasions the INTUC too has taken a strong position on automation or rationalisation in a particular plant or company.

turbojets and pure jets by Air India and Indian Airlines caused considerable organisational upheaval but little labour difficulty or resistance, as such.

Advanced technology in Indian industry frequently has interesting union by-products, as the skilled groups seek to differentiate themselves from the mass of unskilled employees. The aircraft technicians (mainly ground mechanics) formed a separate union to maximise their special strategic bargaining position on the airlines. The crane operators in the Bhilai steel plant (about 100 out of a workforce of almost 12,000) have tended to operate as a small strategic bargaining group within that large complex. Separate groupings are also developing on the railways among locomotive drivers, station-masters and other highly skilled occupations.

To some extent one can discern a degree of differentiation between Indian worker and union resistance to rationalisation, on the one hand, and "pure" automation, on the other. While the mechanisation of manual processes may be cautiously seen as evolutionary and digestible, automation often appears indigestible and almost cataclysmic in its potential efforts. At a special labour-management-government conference called to consider automation, the All-India Trade Union Congress declared: "Whereas the possible redundancy due to simple rationalisation could be set off to an extent against the normal rate of attrition in the labour force and the normal rate of capacity expansion, the redundancy caused by automation involving several times over the impact of rationalisation cannot obviously be contained through any such means." [1]

The Indian National Trade Union Congress, at the same conference, remained so fearful of the country's unemployment problems that it proposed that "a standing compact tripartite committee should be set up to screen all proposals for automation whose clearance should be necessary before any scheme of automation be put into effect". Such a committee could help "remove difficulties and resolve disputes as and when they may arise".[2]

At the same conference again, Indian employers denied the evil employment effects being attributed to automation, and argued that the positive value of increased production and new and higher job skill requirements would outweigh any disadvantages. The employers indicated their willingness to discuss the introduction of automation with the unions, seemed almost ready to guarantee no retrenchment, but balked at taking responsibility for the total employment problem of the society.[3]

---

[1] All-India Trade Union Congress: *Automation in India*, paper prepared for the Indian Standing Labour Conference, 28th Session, New Delhi, 18 July 1968, p. 4.

[2] Indian National Trade Union Congress: *A Note on Automation*, paper prepared for the Indian Standing Labour Conference, 28th Session, New Delhi, 18 July 1968.

[3] See the paper submitted to the same conference by the Employers' Federation of India: *Automation—Blessing or Curse?* Monograph No. 10 (Bombay, 1968).

No agreement was forthcoming at this conference, although there seemed to be some general sympathy with the INTUC's call for a new special screening committee.

In Latin America the level of development is higher than in Asia and Africa and, by and large, collective bargaining is often more sophisticated. Generalisations about unionism and the response to technological change are difficult; but in at least a few instances where an older industry has sought to embark on a major rationalisation programme (as, for instance, in the Bolivian tin mines), fears of job loss and old sentiments of hostility to management have combined to create a major crisis. In the advanced technological sections of most Latin American countries, however (as in the steel, non-ferrous metals or automobile industries), the union response to technological change is tending to take the form of rather modern demands for benefits, protection and participation in job evaluation, incentive setting, and so on.[1]

INFLUENCE OF TECHNOLOGY ON THE ORIGINS AND FORMS OF UNIONISM

The form and character of unionism in any one country reflect, of course, the variety of social, economic and political forces shaping that country's key institutions. In the case of most less developed countries one must also emphasise the important influence of the pre-independence colonial pattern. Students have often contrasted the character and style of unionism in the former British colonies of Africa, on the one hand, with that in the former French colonies on the same continent.[2] Roberts and de Bellecombe, for example, stress the tendency in British Africa for more plant-level bargaining to develop, as against the more national, tripartite, legalistic pattern of negotiations which took root in French Africa. Independence has modified these earlier patterns considerably, but some significant industrial relations differences between these areas can still be traced to colonial influences.[3]

Isolating technological forces, including automation, and assessing their influence on the form and character of unionism in less developed countries is a difficult matter and involves the student in making almost unrealistic,

---

[1] One does find an occasional isolated case of resistance to technological change in Latin America. In 1968, for example, a strike was conducted for more than three days by the stevedores in five Venezuelan ports, as they protested against unloading a new British container ship.

[2] See B. C. Roberts and L. Greyfié de Bellecombe: *Collective Bargaining in Africa* (London, Macmillan; New York, St. Martin's Press, 1967).

[3] ibid., especially Chs. 2 and 4.

artificial distinctions. Keeping this limitation in mind, the following general comments appear relevant.

Nearly everywhere in the earlier industrialising nations of Europe and North America the first unions were organised by skilled craft workers (carpenters, painters, tailors, shoemakers, hat makers, etc.).[1] These unions were generally a reaction against the growth of the capitalist market, which undermined the earlier economic independence of both skilled masters and journeymen. In most instances in the eighteenth and early nineteenth centuries, as capitalism moved (relatively) slowly, first to a commercial and later to more industrial forms, the earliest unions were based on skilled crafts. The leaders were tradesmen themselves—carpenters, printers, and so on. Contrary to some popular misconceptions, it was not the advent of the factory system that produced early Western European (or North American) unionism. Factory unionism came later.

Different economic circumstances led to the emergence of unionism in the less developed countries. (We shall ignore here the very important political forces, as we are concentrating on technological economic influences in this paper. But the reader is reminded that in many parts of Asia and Africa political leaders and political forces—particularly the struggle against colonialism—often played a critical role in unionism from its beginnings.) In most of Africa and Asia there was no slow step-by-step evolution of merchant capitalist forms.[2]

Rather modern industrial forms (mines, railways and even factories) were often introduced by business forces from Western Europe on top of more traditional economies in Africa and Asia. Old handicraft industries were quickly swept away or left in the backwash of the economy. The skilled craft workers were usually in no position to take the lead in establishing unions. The key centres of new economic activity were often these important islands of technological modernism. The late beginnings of industrialism in most of Africa and Asia gave it a special character, and some of these special characteristics carried over to the unionism which emerged.

For example, nearly everywhere one looks in the Asian and African world, the railways—in their way the most advanced forms of technology and economic organisation in late nineteenth- and twentieth-century African and Asian nations—were among the first (often the very first) centres of

[1] The writer has dealt comparatively with this stage of unionism in his book on union development, *Trade Unions and Industrial Relations: An International Comparison* (New York, Random House, 1969).

[2] A few Latin American countries, so far as economic development and unionism are concerned, did undergo a process somewhat similar to that of Western Europe, with craft groups reacting to the growth and pressure of markets and forming the first unions. This, for example, was to some extent the case in Argentina.

enduring unionism. Instances are many: the emergence of an important railway union in India in the 1920s; the early development and great lasting power of the railway workers' union in Ghana, through the colonial administration and the difficult political changes since Independence; Ananaba remarks that "it is often believed that the inauguration of the Railway Workers' Union marked the beginning of militant unionism in Nigeria . . ." and mentions an important rail strike as early as 1921.[1] Scott describes the Uganda Railway African Union as the "oldest" and "most consistently active" in that country.[2] Railway unionism shows relatively great durability in such West African countries as the Ivory Coast, Nigeria and Upper Volta. A study of the origins of unionism in the Sudan suggests the decisive early importance of the struggle by the railway workers to gain recognition for their Workers' Affairs Association, beginning in 1946. This struggle had been preceded by the formation of fraternal clubs of railway workers for social and cultural purposes, as much as a decade earlier.[3] Strong and continuous unionism in Argentina dates back to the founding of the famous railway workers' union La Fraternidad in 1886, and throughout the many vicissitudes and upheavals in Argentine history since then, railway unionism has generally endured and been a main centre of unionism and bargaining in that country.

Not only does African and Asian unionism often have some of its earliest beginnings in railways, but the durability of these unions also attests to their almost unique importance. For example, one also finds early sporadic efforts at plantation unionism in many of the African countries where railway unionism took good root. Here too, foreign capital was generally in the lead. But the efforts to unionise plantations often ended in failure and generally enjoyed much less success than railway unionism. The explanation (allowing for national variations) seems to derive from several factors, some of them technological. The railways were large employers (especially in backward countries) and the concentration of workers makes them a more likely target for successful unionism; however, to some degree the same can be said of plantations. But the railways generally attracted and were dependent upon a high level of skill because of the high level of technology involved. This skill factor doubtless carried with it a potential element of bargaining power (often lacking in plantations)—and it also entailed above-average literacy.

---

[1] Wogu Ananaba: *The Trade Union Movement in Nigeria* (London, C. Hurst & Co., 1969). References to railway unionism and its importance are scattered throughout this volume.

[2] Roger Scott: *The Development of Trade Unions in Uganda* (Nairobi, East African Publishing House, 1966), p. 59.

[3] cf. Saad Ed Din Fawzi: *The Labour Movement in the Sudan* (London, Oxford University Press, 1957), especially Chs. 2, 3 and 4.

One cannot exaggerate the factors of skill and literacy in understanding the development of unions in adverse surroundings. Indeed, in looking at the first leaders of railway unions one finds that they are often individuals well up in the white-collar and technical ranks. It would also appear that the very nature of railway work and organisation, which brought workers into contact with a wide variety of different geographical areas, including important cities and ports, helped to stimulate or provided a more favourable setting for unionism.

While it is hard to estimate the extent of its importance, the fact that the governments owned or controlled the railways in most African and Asian colonies may have enhanced the possibilities of union recognition, at least in the twentieth century when more liberal colonial labour policies were being developed.

Not only was railway unionism to prove more durable than many other forms in Asia and Africa, but its policies and practices appeared to be more realistic and more geared to the economic and social needs of its members than was the case of other unions. At an early stage one often found railway unions insisting on written collective agreements and concrete gains in wages, hours and working conditions.

National or so-called category unions, based on either an industry or a craft, have been relatively slow to emerge in many less developed countries.[1] The paucity of such category unions usually contrasts with the plethora of unions with a geographical (limited to a city or small region), general worker, or single enterprise base. The railway unions, of course, tend to be an exception. They operate in a continuous highly interrelated and interdependent market situation and this makes it a virtual necessity for them to strive, from their beginnings, for broad geographical (usually national) coverage. The economic and technological organisation of the railways themselves is decisive in influencing the forms of railway unionism. National unionism may also be as much a necessity for railway management as for unions. That is not so for many other industries and enterprises. Such enterprises may often be unique (without serious competitors) and isolated from market pressures or needs. Here, single enterprise unions may feel little need to federate into a national union. It is not surprising that category unions are slower to develop in many other industries.[2]

---

[1] This is in contrast to the formation of national confederations, which unite unions (regardless of industry or region) at the all-country level and often reflect political pressures.

[2] In India even the relatively durable unions of the textile workers in Ahmedabad and the steel workers at Tata (Jamshedpur) are almost fully independent, with only slight, and as yet not very critical, ties to any national or category unions. They do belong to the latter, but carry on the great bulk of their work independently, in their own areas.

As industrialisation proceeds, however, and as markets become more national in scope, one can expect further development of category or national unions. In the new countries these generally tend to be along industrial lines, particularly for blue-collar workers. Top governmental leaders, politicans and economic planners usually do not look kindly upon craft unions, which are likely to be more difficult to integrate into national wage and planning structures. Governmental intervention in union affairs, to bring about a more compatible trade union structure, has become fairly common in Africa during the past decade.

IMPACT ON COLLECTIVE BARGAINING AT THE ENTERPRISE LEVEL

The firms in the advanced sectors of less developed countries often provide a favourable setting for more "advanced" industrial relations patterns and practices. This favourable atmosphere may be due to a combination of technology, organisation and politics. The technological-managerial problems in a modern steel plant, automobile factory or petroleum refinery call for a more sophisticated managerial response than is required elsewhere in the less developed world. Large investments in modern plant must be carefully managed, if expensive equipment is not to be idled as a result of labour difficulties. Often, too, this management has been (or is) subjected to some foreign (usually European or American) influence, either because the plant has been built with the help of some advanced country's design or technology, or because the plant itself is a branch of some foreign firm.

Because of these factors, the managers of such firms frequently have had (or have) some contact, direct or indirect, with industrial relations practice and unionism in a more developed country. Confronted with union demands for recognition and negotiations, such managers are less likely to react with traditional ideological hostility than might otherwise be the case of an early-industrialising managerial élite. These advanced-type plant-enterprise managers live, so to speak, in both the developed and less developed worlds.

By way of contrast, for example, observers have often found that Chinese owners and managers of mines and other enterprises (usually smaller in scale than their European counterparts) in Malaya react far more negatively to unionism than do the other managers (and owners), usually British, with developed-country experience and/or relationships. In Africa too, it appears that unionism is far more easily established in large foreign-owned and/or managed mines, docks and factories than in other industries which are managed and owned by the country's own nationals. In part, of course, the acceptance of unionism in such advanced sectors as mines, refineries and factories may also be attributable to the feeling of political vulnerability on the part of foreign firms operating these facilities in former colonial countries. Brusque treatment of the union on the part of a "foreign" management may lead the workers to seek redress for their problems through politico-governmental channels.

The personnel recruitment practices of the more advanced and larger firms may also enhance the possibilities for unionism. While it is difficult to generalise, there is a tendency for the more advanced (i.e. advanced technologically, organisationally, in size, etc.) firms to recruit a more experienced workforce. A new steel or auto firm, paying higher wages than most of the remainder of the economy, often with more careful selection policies, is more likely to select a worker who is already urbanised, or has at least made some clear break with the countryside. Often this may be the worker's second or third industrial job—and in some instances he may even be a second-generation urban dweller (that is, one whose father migrated earlier from the countryside).

Workers such as these are likely to be more committed to an industrial way of life. They are likely to be more literate than the population average. In turn, these characteristics are likely to enhance the appeal of unionism, particularly as they become caught up in the bureaucratic coils of a modern economic enterprise. Union leadership is also easier to produce among such workers than in less permanent, less bureaucratic, more personal structures which may characterise other firms in a less developed country.

Related to these factors is the growing internationalisation of the labour-union response to the activities and operations of certain industries and corporations. In the petroleum industry (the most advanced technological sector in a number of countries, most notably in the Middle East), the influence of international trade union meetings and contacts is clearly evident. Petroleum workers in less developed countries today easily establish contact with those in developed areas, and soon confront their managers at the bargaining table with demands for many of the same benefits as are already established elsewhere. Check-off, union participation in rate setting, shift premiums, and other features of developed-country bargaining, begin to crop up in petroleum refining collective agreements in many less developed countries.

Latin America, at a considerably more advanced stage of industrial development, provides vivid examples of the influence of so-called multinational (trans-national) companies, on the one hand, and international trade union secretariats on the other.[1] An examination, for example, of automobile company agreements in Latin America shows a surprising number of the provisions found in European and North American agreements. The same is true of the agreements covering steelworkers in many parts of Latin America.

In the steel agreements in Latin America, for example, one frequently finds provision for a joint union/management committee to deal with the problems of managing any job evaluation plans (a practice which began in the steel industry in the United States some 25 years ago). Union/management committees to consider production standards are fairly common in steel and automobiles.

While some of these practices may reflect a carry-over of company practice from North American or European plants, they also spring from the co-operative exchange of data (via bulletins, joint regional conferences and international conferences) and exchange visits among union officials or unionised workers employed by these same companies, in different parts of the world.

It is increasingly common in parts of Latin America and the Caribbean area to find a young worker quite stimulated by the challenge of working for a world-wide aluminium trust or automobile corporation. He responds with an intensive study of "his" company and its labour and social policies elsewhere, often with the result that his own union comes to demand the most advanced practices and benefits. In this work the unions in the less developed countries are frequently assisted by visiting bargaining and technical experts from the more advanced nations. These experts may be provided by international trade union secretariats or on a bilateral basis by the union in the advanced country.

The export of union expertise more and more follows the export of advanced technology and management, especially in capital-intensive industries which generally offer greater bargaining possibilities.

WHITE-COLLAR UNIONS AND AUTOMATION

Some of the few specific cases of union reaction to automation at the enterprise level in the less developed countries have occurred at the white-

[1] The publications of the International Metalworkers' Federation, a trade union secretariat based in Geneva, have been extremely useful with regard to collective agreements in practice in all branches of the metals industry in Europe, North America and Latin America. The writer has also drawn heavily on conversations with officials of the IMF in preparing the next few paragraphs of this paper. The trade union secretariats are groupings of unions on an intercountry basis, along the lines of branches of industry. Thus, there are secretariats combining the world's transport unions, petroleum unions, public service unions, metalworkers' unions, etc.

collar work area.[1] Several years ago, for example, the Chilean steel company CAP at Huachipato began to computerise its white-collar operations. The company appeared to be concerned with its mushrooming white-collar workforce, in a period when the number of blue-collar workers had been nearly stationary. The introduction of the computer led to some lay-offs among white-collar employees, and threatened others. The white-collar union (Chilean law provides for separate unions of blue- and white-collar workers) reacted vigorously and eventually bargained through a special agreement which provided that: there would be no lay-offs among existing personnel as a result of computerisation; "natural" attrition would be the source of any employment reduction; to the maximum extent possible, workers already employed were to be trained for the new computer positions; employees who had to be shifted because of the impact of the computer were to suffer no loss in pay. Those who had been laid off were rehired.

In a somewhat similar situation in 1967 in Guyana involving the introduction of computerised operations into the offices of a major bauxite producing company (ALCAN), the miners' union (which represents both blue- and white-collar workers) negotiated an agreement, similar to that in the Chilean case. It provided for reliance on attrition as a means of reducing the clerical labour force, retraining rights for workers, and so on.

Confronted with the installation of high-speed modern office equipment in some commercial banks of India, the All-India Bank Employees Association entered into a bipartite agreement with the bank employers' association, in October 1966, which included a "chapter" on mechanisation. This chapter accepted management's right to introduce a variety of office machines (including National Cash Register, Remington Rand and Blue Star accounting machines, etc.). The agreement stipulates there is to be "no retrenchment . . . and the displacement of staff in a particular department or office/branch where such machines are introduced will be kept at the minimum possible level". (Presumably displaced workers must be absorbed elsewhere in the enterprise.)

The reaction of various unions to the proposed computerisation of some operations in the Indian Government's Life Insurance Corporation has, however, been far more violent. Here, despite government assurances that no retrenchment would occur, union opposition has continued. The union counter-argument, frequently heard in this conflict, has been that computerisation will undermine *future* potential new employment opportunities and therefore should not be allowed in a country with India's population problems.

[1] This should not be thought too surprising. While there are not too many instances of the automation of blue-collar work in less developed countries, in a fairly substantial number of instances some white-collar work has been automated.

Countries which are confronted with the special problem of unemployment among well-educated sections of the labour force are likely to be highly sensitive to office-type automation. The Indian National Trade Union Congress calls for special "scrutiny in the case of automation of 'table work' which will displace clerical hands and to the unemployment of the educated which is extremely dangerous".[1]

It is likely that office mechanisation and computerisation will continue to be a source of potential difficulty in India and other less developed countries. White-collar workers often have less experience with modern mechanical processes than blue-collar workers, and may be less able to cope with their introduction. Then, too, white-collar workers may be more status-conscious about the particular jobs they hold, and, consequently, resistant to major changes, even if their employment as opposed to their particular job status is not threatened. Thus, despite reassurances that there will be no individual retrenchment, office unions are likely to resist automation, and at least try to slow down its introduction.

* * *

Union experience with automation as such in the less developed countries has as yet been relatively modest. This is likely to change in the years ahead as the pace of development increases and different countries (and enterprises) attempt to leapfrog ahead technologically. Some of the geographical differences in the reaction to technology, sketched in the first part of the paper, are likely to persist in the years immediately ahead, and those areas or countries which have already experienced some resistance to major technological change and automation are likely to continue to do so.

---

[1] *The Indian Worker* (New Delhi), 22 July 1968, p. 3.

CASE STUDIES

COMPUTERISED BANKING IN BRAZIL [1]

CELSO ALBANO COSTA
Banco do Brasil SA, Rio de Janeiro

Banco do Brasil SA was founded on 12 October 1808 as a government bank, and started operations at the end of 1809. The bank has functioned continuously since then, except for the period 1838-54, when it was closed. It is the largest Latin American bank, with total assets of about US$ 6,584 million at the end of 1969. It plays a prominent role as a federal agency but does not, however, act as the Central Bank (although it once resembled such an institution). It handles public and other accounts in a normal banking manner. At the end of 1969 it had 684 branches in Brazil, 23 more than in 1968.

During the period 1940-69 the number of employees increased almost tenfold—from some 4,500 in the earlier year to about 11,800 in 1950, approximately 25,700 in 1960 and almost 42,500 in 1969. Table 4 traces the slow growth of employment in the most recent years, during which time the number of branch units increased by 101, or more than 15 per cent.

Until 1962 the bank had only one employment register for employees, with no distribution between the General Office and the branches. Table 5 shows the distribution of employees between the General Office and the branches, and provides a further breakdown showing the total numbers of employees and clerks, for selected years after 1962.

This paper is especially concerned with the "clerks" category, which includes employees serving as managers of the branches. As shown in table 5, their share of total employment levelled off after 1966. Nevertheless, there was a shortage of 1,500 clerks in December 1969.

[1] This paper, for which the writer has sole responsibility, was made possible through the courtesy of the President of Banco do Brasil SA, Dr. Nestor Jost, to whom the writer is most grateful. Thanks are also due to the writer's colleagues, especially those from the departments discussed. Without their co-operation it would have been impossible for him to perform his task.

Table 4. Growth of branch units and total employment, Banco do Brasil SA, 1964-69

Year	Number of branches	Percentage increase over previous year	Number of employees	Percentage increase over previous year
1964	583	—	38 448	—
1965	629	8.0	39 395	2.5
1966	645	2.6	41 650	5.6
1967	650	0.8	41 699	(+)
1968	661	1.7	41 703	(+)
1969	684	3.5	42 457	1.8

Table 5. Employment characteristics, Banco do Brasil SA, for selected years after 1962

Year	Number of employees	Number of clerks	Distribution of employees	
			General Office	Branches
1962	31 162	21 161	5 281	24 254
1966	41 650	31 488	6 712	34 938
1969	42 457	31 582	7 197	35 260

Most people working at the Central Bank are also employees of Banco do Brasil SA. After 10 years they can exercise the option of staying permanently with the Central Bank. About 4 per cent of the employees represent people working in other similar governmental departments.

FROM MECHANISATION TO COMPUTERS

The first study of mechanisation at Banco do Brasil SA was made in 1940, but the bank did not mechanise until 1945, when accounting machines (National) were bought for the São Paulo branch.

In view of the good results obtained from mechanisation at this branch, management concluded it would be advisable to extend mechanisation to other branches and also to standardise and centralise its various services. This programme was started with IBM punchcards in Rio de Janeiro in 1953.

The General Mechanisation Service was created in 1956 and soon raised to departmental level, just one degree below directorial level; it was renamed the Mechanisation and Telecommunication Department (DEMET).

DEMET officials realised the need for personnel who were sufficiently competent to keep up with new data processing methods, and were helped in this by the creation of the Personnel Selection and Development, General

Table 6. Classification of branches, Banco do Brasil SA, by type of mechanisation, 1967 and 1969

Group	Type of mechanisation	Number of branches	
		1967	1969
1	None	323	64
2 (Stage A)	Entitler set (Addressograph) Teller adding machine (Burroughs 10-10-360 and 10-10-382) Accounting machine (Burroughs P-623-B)	108	298
3 (Stage B)	Same as Group 2, plus the accounting machine National 31-A	173	234
4 (Stage C)	Electronic data processing (IBM 1401; Burroughs 3500; Bull-Gamma 30; calculators Bull-Gamma 3; IBM 360/40) [1]	41	88

[1] Stage C has the following additional equipment: 2 punchcard machines (IBM 026); 1 paper tape punch (Olivetti 1733), with check digit; 1 paper tape punch, without check digit.

Department (DESED), set up in 1965 to increase the know-how of the employees.

Thus, two large areas were opened up within the framework of mechanisation: (*a*) that of the accounting machines, with the training programme under the responsibility of the new department (DESED); (*b*) automation, electronic data processing, and a training programme carried on by IBM. Computers were imported at reduced tariff rates.

Although the bank had had a punchcard system operating in Rio de Janeiro since 1953, as we have seen, and had installed four electronic data processing centres more recently, most of the branches did not have any kind of mechanisation until 1967. A general plan was drawn up in 1966 for the mechanisation of all the branches according to their individual possibilities.

After the establishment of the electronic data processing centres and the personnel selection and training departments, the next important stage was to be the rationalisation and standardisation of services in order to get the maximum from automation. To do the job, the Mechanisation and Telecommunication Department was transformed into the Organisation of Services and Communication, General Department (DESEC). Thus the automation, rationalisation and standardisation of services were under one head.

Under the 1966 mechanisation plan, all branches were classified into four groups. Table 6 shows the position in 1967 and 1969.

To utilise the capacity of the computer centres as much as possible, the department has organised "preparing centres" which carry out the intermediary work of obtaining the records, punching them and sending them

Table 7. Training programme, Banco do Brasil SA

Subjects	Hours
Branch service organisation	29.5
Telecommunication	5.0
Statistics (introduction)	10.0
Management (introduction)	6.25
Training techniques	7.5
Human relations	12.5
Accounting machines—National 31	41.0
Accounting machines—Burroughs P-623	25.5
Teller adding machines	4.0
Addressograph	4.0
Organisation	13.5
Total	158.75

to be computerised. The bank has 12 such centres, grouping 138 branches; each centre serves the surrounding branches within a radius of 100 km.

Each of the branches within that radius accumulates the daily transactions and sends the documents by messenger to the preparing centres, where the data are coded and transferred to tapes, which are sent to the computer centre to be processed. The computer produces a report to be sent back to the branches in time for the next day's operations.

PERSONNEL DEVELOPMENT

At the earliest stage of mechanisation, the Accounts Section was in charge of the training programme. National or Burroughs had provided insufficient technical assistance to satisfy the demands of the bank; bank staff thus had to complete the task.

At present a large group of candidates takes the first examination. DESEC then sends them to DESED for a psychological test. At that stage there are still four candidates for each vacancy. The training programme, of 28 days (full time) for 50 trainees, is shown in table 7. After the course, DESEC keeps the best 25 trainees working at the General Office, while the other 25 go back to their branches.

Submanagers' training

Every submanager is responsible for mechanisation at his branch. For this purpose all submanagers are trained in Rio de Janeiro by DESED.

Table 8. Submanagers' training programme, Banco do Brasil SA

Subjects	Hours
Business law	12.5
Branch service organisation	20.0
System organisation	12.5
Human relations	12.5
Telecommunication	3.75
Training techniques	7.5
Teller adding machines	7.5
Lay-out	3.75
Addressograph	7.5
Accounting machines—Burroughs P.623	18.75
Routines	12.5
Total	118.75

The main goal of DESED was to meet the needs of all branches by August 1970. In 1967, 237 submanagers were trained; the figure went up to 490 at the end of 1969, when 81 remained to be trained to complete the programme. The need since then has been to keep the submanagers' knowledge up to date.

The subjects treated during the training programme (20 days, full time) are partly theoretical but mostly practical, as shown in table 8.

Programmers' training

In recruiting employees from several branches and departments, the psychological test was used in the first phase of selection. This test, initially under the responsibility of a private institution for professional guidance, is now administered by DESED.

The training programme itself was carried on by IBM for 36 candidates. Of these, 25 were nominated as programmers after 90 days. Table 9 shows the magnitude of the training programme developed and sponsored by Banco do Brasil SA from 1965 to 1969.

CHANGES IN OPERATIONS

Production methods were completely changed, the flow of services was technically determined, a scientific basis was adopted to calculate the productivity indices, and finally the computer was introduced to help management

Table 9. Magnitude of the training programme sponsored by Banco do Brasil SA, 1965-69

Trainees	1965	1966	1967	1968	1969	Total
Tellers	—	167	1 542	1 105	2 183	4 997
Submanagers	47	87	103	199	54	490
Managers	—	—	6	248	202	456
Inspectors	—	—	22	100	42	164
Assessors and assistants	—	—	32	45	63	140
Electronic data processors	36	— [1]	— [1]	— [1]	— [1]	36
Others	33	256	610	1 913	826	3 638
Total	116	510	2 351	3 610	3 370	9 921

[1] Sponsored by IBM during 1966-69 and therefore omitted.

take decisions by obtaining information from the branches and automatically to determine the number of employees needed for each branch.

We shall illustrate the changes in production methods by three examples. The first case will examine the modifications in the methods of selecting candidates to be hired, the second will present the payroll sheet for all the General Office employees, while the third will show the changes in services to clients.

Selection methods, DESED

Banco do Brasil SA appoints its staff on the strength of the results of a regional or nation-wide public competition.

The older system prescribed essay examinations in all relevant subjects. All examinations were graded by some 60 teachers and instructors over a period of two months. The whole job was done by hand with typewriters and adding machines, so it would be impossible to obtain statistical interpretations or make scientific observations.

The new system presents two very significant changes: (*a*) the introduction of a psychological test; (*b*) objective multiple-choice questions with an answer sheet.

In November 1969 the bank had a second chance to select people according to the new system. This did not gain very much time, but it did save on salaries. The bank dispensed with about 60 teachers and instructors for the grading; instead, only 20 people were needed to punch and check cards. (The teachers had not organised, and did not protest.)

To complete the change the bank must try to reduce the time between the examinations and the final decision and is doing so experimentally with a reader machine (IBM) to eliminate the punchcard operation. As soon as good results are obtained from that equipment it will be possible to eliminate some steps in the process.

It is necessary to stress that cost control became easier with computerisation, because all the expenses of selection are classified, enabling the cost of each enrolled candidate to be calculated. The figure for each enrolled candidate was NCr$ 24.70 (about US$ 5.80) and for each selected candidate, NCr$ 395.32 (about US$ 94.00).[1]

Payroll

The General Office payroll, our second example, will be discussed in terms of the whole system, which has already been approved but only partially adopted. The most economic method of mechanisation would require a simplification of flows, with some jobs being standardised and with centralised control and processing.

At an earlier stage, the General Office payroll had been under the responsibility of the Accounting Department, within which there were two sections, one for attendance control and the other for the payroll. Due to the bank's policy of making no deduction per hour or halfday, payroll computation is relatively simple.

In 1966 the responsibility for the job was transferred to the General Personnel Department (FUNCI), and grouped within a single section. This change brought a high degree of simplification permitting a reduction in staff from 106 to 80; 60 others were no longer needed to check the data related to 6,900 employees on the General Office payroll, and these 60 were all absorbed into other offices of the bank in Rio.

The first step was the introduction of microfilm to simplify the filing. The job was done for every General Office employee, covering the data from 1963 to 1966 and making a total of 420,000 sheets. Sheets are photographically processed soon after leaving the computer. Thus space is reduced and intermediary steps are eliminated.

All payroll deductions were classified, including loans from federal organisations, and entered on the payroll sheet monthly. Salary advances were grouped into two classes according to the time agreed for reimbursement (10 or 25 months) to facilitate control by magnetic tape, and they were all contained in one ledger account rather than three. As a result FUNCI could eliminate thousands of typed cards and sheets, with a corresponding saving on storage space and insurance.

---

[1] US$ 1.00 = NCr$ 4.40.

Some idea of the practical consequences of this rationalisation may be drawn from the fact that 60,000 cards were reduced to 8,000.

The final stage will be the introduction of automatic computation of periodical promotions, covering wage increases, the five-year over-salary premium, vacations and other regular occurrences.

Service to clients

The creation of the Central Bank involved the transfer of some jobs from Banco do Brasil SA, which had to move towards banking services in general, competing with several Brazilian and foreign banks. Entry into these new markets was accompanied by a steep increase in competition. The Government's policy of a forced reduction in the number of loans entailed efforts by all banks to reduce costs. In order to compete, Banco do Brasil SA has improved its services. The computer offers a better service to clients: improvement of quality is evident and is partly responsible for the increasing number of new clients.

The new system introduced by Banco do Brasil SA was based on the experiences of North American banks, where the tellers pay cheques and receive deposits direct from the clients with no intermediary employee being involved in the transaction. This was made possible by a training programme which prepared the tellers (or "executive cashiers") for the new system, as part of which they would need to be able to recognise signatures and understand the routines and the role of the bank within the economic context. Documents are processed by computer; the tellers themselves, however, need no computer skills.

DECISION MAKING AND PRODUCTIVITY

In order to develop the use of computer facilities, Banco do Brasil SA decided to study the possibility of introducing a productivity index, a standard measure to evaluate all branches. This index was determined on the basis of data from a series of periods, and the branches have now automatically adjusted the size of their staff to coincide with the standard number shown by the index. Until 1968 the standard was based on a sampling of 30 branches with data going back to 1954. The new system takes into consideration data on all the branches, grouping them in four classes according to their degree of mechanisation (see table 6).

The average productivity index, calculated according to a complex system, was found to be 0.81, and decisions about the number of employees per branch were based on the following principles:

— no change for branches with normal productivity, or low productivity of not less than 0.51; reduction in number of employees for branches with productivity of less than 0.51;
— an increase in the number of workers for those branches with a productivity index equal to or more than 0.81 unless one additional employee would bring the index below 0.81.

The final result was an increase of 172 employees.

The good results obtained from the new policy are confirmed by a jump in the average index from 0.81 to 0.98, and a rise from 59.5 to 69.4 in the percentage of those branches having normal, high and very high productivity.

Checking accounts per employee

The relationship between the number of employees and the number of checking accounts was determined by sampling all the mechanisation centres. The sample was based on the following criteria: (*a*) only checking accounts (voluntary deposits) were included; (*b*) the period was limited to 1962 to 1969, since Banco do Brasil SA did not distinguish between accounting clerks and others (lawyers, engineers, and so on) before 1962.

Table 10 shows the results obtained. It should be noted, however, that relative productivity cannot be determined from the table, as employees have other tasks that do not vary between cities in proportion to the number of checking accounts.

SALARIES

The labour market for electronic data processing is highly competitive in Brazil, one reason being that IBM, Burroughs and other computer firms sold more machines than they trained people to work them.

Banco do Brasil SA has some advantages within this market. It is first as regards wage levels (e.g. a salary year of 16 months) and provides fringe benefits such as very low life-insurance premiums, medical care, retirements benefits, insurance and loans to buy houses. Employees are granted automatic promotion until they have been working for the bank for seven years, after which time promotion is determined by merit and length of service.

The employee receives a premium according to his position in the hierarchy. There is a direct relation between the classification of an employee

Table 10. Mechanisation system, accounts and employees, at four branches of Banco do Brasil SA, 1962-69 (end-of-year data)

Year	Type of mechanisation	System	Number of accounts	Number of employees	(4:5)
Branch: Belo Horizonte					
1962	Accounting machines	Conventional	7 823	334	23.5
1963	,,	,,	9 277	365	25.5
1964	,,	,,	9 812	365	26.8
1965	,,	,,	10 407	367	28.5
1966	IBM Bureau	,,	10 959	367	29.5
1967	,,	,,	9 628	317	30.2
1968	,,	Teller/Nov.	11 151	314	36.0
1969	,,	,,	17 955	337	53.0
Branch: Brasilia					
1962	Bull-Gamma 3B	Conventional	11 582	236	49.0
1963	,,	,,	14 827	249	59.6
1964	,,	,,	22 103	258	85.5
1965	,,	,,	19 958	273	72.0
1966	,,	,,	23 742	288	82.1
1967	,,	,,	38 223	362	105.8
1968	,,	Teller/Sep.	43 930	337	130.0
1969	,,	,,	49 345	383	128.2
Branch: Recife					
1962	Accounting machines	Conventional	5 236	280	18.7
1963	,,	,,	5 650	315	17.9
1964	,,	,,	6 343	333	19.2
1965	,,	,,	6 826	330	20.7
1966	IBM Bureau	,,	7 028	337	20.7
1967	,,	,,	19 109	325	58.8
1968	,,	Teller/Sep.	24 287	325	74.5
1969	,,	,,	34 493	324	106.0
Branch: São Paulo					
1962	Accounting machines	Conventional	36 897	1 437	26.2
1963	Bull-Gamma 30A	,,	41 674	1 453	28.9
1964	Bull-Gamma 30A and 30B	,,	44 139	1 642	26.9
1965	,,	,,	51 262	1 750	29.3
1966	,,	,,	59 373	1 797	33.0
1967	,,	Teller/Nov.	59 072	1 533	38.5
1968	,,	Teller	59 824	1 528	39.0
1969	,,	,,	71 168	1 715	41.6

and the positions he can occupy, which sets a traditional limitation to the income of the new employee. However, the tradition is not observed for trainees, analysts, programmers and some others, so that those people have the possibility of receiving 40 or 50 per cent over the fixed wage.

Let us suppose that a systems analyst with a fixed wage of NCr$ 1,500.00 monthly receives a premium, related to his function, of NCr$ 648.00. Since, as we have seen, 16 payments are made annually, he receives a premium of NCr$ 864.00 per month. In dollar terms this is around US$ 192.00 monthly.

Table 11. Average monthly premium paid to electronic data processors, Banco do Brasil SA

Function	Average monthly premium (in US$)
Division head	286.00
Head of mechanisation centre	238.00
Head of programmers	216.00
Analyst, supervisor	192.00
Technical assistant, sectorial head, programmer	166.00
Technical sub-assistant, management assistant, programme assistant	120.00
Management and technical auxiliary	86.00
Head monitor	44.00
Monitor	40.00
Printing operator	33.00

The average monthly premium paid to electronic data processors, according to their function, is shown in table 11.

Annual wage rises equal the annual percentage increase for all bank employees.

CONCLUSIONS

Banco do Brasil SA, as the largest Latin American bank, was obliged to improve the quality of services it offered its clients. After several stages of mechanisation, computerisation was introduced in the main branches—although there are still 64 branches with no mechanisation at all, since mechanisation has its own difficult training problems and requires electricity which is lacking in some remote towns. Computerisation is more useful for repetitive work, and (as at many other banks) the checking account department was the first to be automated, due to the high volume of work. Payroll accounts came next.

The decision to adopt automation was taken by the bank management with no adverse reactions from employees or unions, because it improved the wages of some people without any negative effects on the employment level. Due to this situation and the morale of the employees, there was no conflict. Moreover, opposition to computers has not been characteristic of Brazil anywhere.

One of the challenges of automation is the danger it might represent for the workers by provoking unemployment. Banco do Brasil SA did not discharge any employee because of automation—not even those whose jobs were eliminated, since these workers were transferred to other positions created as a result of the bank's growth (opening new branches, offering new services such as guaranteed cheques, travellers' cheques, collecting taxes, etc.). Transferred workers received no bonus or other special compensation.

The bank was able to fill its needs for computer personnel exclusively from people already employed by the bank. As a result, there was an upgrading of the skill level of the employees, not only for those workers at the computer centres but also for those displaced by the new techniques who were retrained to improve their skills, and for those trained for jobs related to electronic data processing.

An increase in output per worker is an obvious result of automation. Output rose significantly as a result not only of the rapidity of computer processing but also of rationalisation, simplification, and control of services (see table 10). As a consequence the bank reduced its unit operational cost.

However, it must be stressed that the combination of rising output and lower costs was not merely a function of automation and better systems but also a result of personnel development. With such development, with above-average salary increases, and with employment security, any large Brazilian organisation can introduce automation or computers. Whether all should do so simultaneously, thus lowering the over-all employment growth rate, is another matter.

MASS PRODUCTION OF CAKES IN COLOMBIA

DANIEL SCHLESINGER
LUCÍA de SCHLESINGER
University of the Andes, Bogotá

ECONOMIC SETTING

Colombian manufacturing in 1966 (the most recently published estimate) comprised 11,800 firms, 63 per cent of which employed nine persons or less. The average number of employees per firm was 25.3. However, firms occupying 100 people or more produced 73 per cent of the total value added in manufacturing, and employed 54 per cent of the total labour force in the sector.[1] Their productivity was higher than the general average for manufacturing.

According to the 1966 estimate, the food industry comprised 3,042 registered firms, 26 per cent of the total number of registered firms in manufacturing. It employed 15 per cent of the total labour hired by the sector (roughly 300,000). The value of its production accounted for 26 per cent and its value added for 16 per cent. Bread and bakery products contributed about one-tenth of the total value of production, and one-eighth of the total value added by the food industry (these fractions are taken from the 1961 manufacturing census).

The production of bread and bakery products is widely dispersed throughout the country, and it is difficult to specify the number of firms having a real industrial operation. Presumably they are more than 2,000. However, there are fewer than 30 firms employing 50 persons or more.[2] In general, this group is characterised by the lowest employment per firm (under 10). Mechanisation is minimal and technology is old-fashioned. Considering the outstanding growth of the food industry in the past decade, this group has merely kept pace with the growth of population.

---

[1] DANE-ANDI: *Guía Industrial de Colombia, 1969,* 1st ed. (Medellín, Publicaciones Técnicas, 1969), pp. x-xvii.

[2] ibid.

Included in this category are establishments producing wheat, corn and yucca breads, biscuits, pastries, cakes, potato chips and cones for ice-cream. In 1966 the production of wheat bread accounted for more than half the total for the group, while the production of biscuits represented around 30 per cent of the total.[1] Cake production was only 5 per cent. However, it must be remembered that the total bread and biscuit production was actually larger, as a substantial amount is made in non-registered small family enterprises with fewer than five persons each.

COMPANY BACKGROUND

The firm Productos Ramo SA is located in Mosquera, a small village in the south-western part of the savannah of Bogotá, about 20 km from Bogotá. The firm is engaged in the production of traditional cakes and (a recent development) of traditional biscuits. Sales in 1969 amounted to 25 million Columbian pesos.[2] The firm employs 231 persons: 197 unskilled workers and 34 technicians and white-collar employees.

According to the *Guía Industrial de Colombia, 1969*, Productos Ramo SA was the largest firm among "cake and pastry manufacturers", one of the 50 categories into which the food industry is divided. However, in general terms, competitors are not only those listed under the same heading but also the relatively large firms manufacturing mainly wheat bread and biscuits, and producing cakes as a side-line. The survey includes about 20 large companies, seven of which are located in Bogotá.

The firm began in 1953 as a family operation with only three employees. The manager, his wife and a maid worked in the kitchen of their home. Each one performed many different tasks: the manager was at the same time director of production, salesman and biller; the wife and the maid were in charge of production and packaging. Procedures were naturally manual. The recipe was that of the traditional family cake of the manager's mother, and the size and shape of the cake was determined by the available pans.

The increase in sales made it necessary to employ more people. The manager added a new storey to his home to house the additional maids who lived there (as many as 20). He also bought a new oven and a mixer. Through a short-term commercial bank loan he was able to secure some

[1] DANE: *Industrial Manufacturera Nacional, 1966* (Bogotá, 1969).

[2] US$ 1.00 = 18 Colombian pesos (Feb. 1970 rate).

transportation equipment, and he hired two drivers. This was the second stage in the firm's development.

In the third stage a legal partnership was established in 1959 with two partners and a capital of 50,000 pesos. By 1961 the first amendment to the original instrument increased the capital to 120,000 pesos, and a year later to 1.2 million pesos. In 1963 the firm moved to another site located in an industrial sector of Bogotá. It became an industrial operation, and its organisation chart was virtually set in its present pattern.

Legal structure

The firm is a close family corporation now having 10 stockholders and a capital stock of 5 million pesos. Fixed assets amount to approximately 36 million pesos. In 1964 the sales and distribution department was formed into a subsidiary firm, Comercial Ramo Ltda., a partnership of 10 members. The subsidiary buys the production and is in charge of selling and distributing it. In 1968, after an unsuccessful attempt to set up an agency to supply Medellín (the second city of Colombia), it was decided to establish a subsidiary plant in that region having a different legal structure from Productos Ramo SA: this was Ramo de Antioquía Ltda. Only the firms covering the Bogotá area (that is, Productos Ramo SA and Comercial Ramo Ltda.) are considered in this case study.

Organisation

The firm is organised under a general manager who is the main stockholder and promoter of the firm. He is a truly entrepreneurial person, being generally regarded as one of the most dynamic executives in private industry.

The board of directors meets regularly, usually to advise the general manager on policy changes and future plans. It is composed of five members, all well-known executives in private finance and industry. The manager's idea is to have a group to help solve the problems of the company instead of a passive group not really interested in its development.

The manager is also advised by five specialists in the fields of administration, mechanical engineering, food technology, labour relations and taxation. According to the manager, "decisions should be decentralised and top executives should be responsible for them"; but the specialists believe he has not followed this precept sufficiently. For example, workers' complaints are made to the director of industrial relations; but if the employee is not satisfied with the solution, he may talk directly to the general manager.

ECONOMIC PERFORMANCE

Market and distribution

Of the total 1969 sales of 25 million pesos, 62 per cent were accounted for by the traditional product of the firm: the Ramo cake. This is a round cake 15 cm in diameter, with a maximum height of 7 cm and a weight of 270 g. Most of them are cut into 12 triangular pieces. The firm also produces other cakes, using the same basic formula plus minor ingredients, such as fruits, chocolate, syrup and artificial flavours. They are named Ramito, Mis Nueves and Gala cakes.

The company is now selling 38 products (considering as one product the various forms into which the basic dough of the traditional Ramo cake is shaped, flavoured and presented for sale.) There are three baking shapes, nine flavours (coconut, zebra, natural, black, chocolate, lemon, fruit, vanilla and orange), two traditional Ramo cake selling presentations (cut and uncut), eight Ramito selling presentations, and *hojuelas* (for jelly rolls).

Sales estimates for the coming year are prepared annually in November and take into account past sales performance of each product plus the expected sales of new products. The sales forecasting for 1970 showed that about 5.5 million units of the traditional Ramo cake were expected to be sold, accounting for 18.6 million pesos out of a total of 30 million.

A basic feature of the success of the company has been the idea of selling the cakes by pieces. Although the firm itself does not sell the traditional Ramo cake by individual pieces, most retailers do so. In this way the product reaches consumers having a very low income level. An estimated 70 per cent of the sales of the traditional Ramo cake are for individual pieces, generally to go with beer and soft drinks. The additions to the product line have had the effect of appealing to other consumer groups having higher incomes and more refined tastes. These non-traditional products are also sold packaged by individual slices. In 1969 slices accounted for 16 per cent of the total sales, while the non-traditional cakes sold as a whole (Mis Nueves and Gala) represented only 8 per cent.

A third type of product is Ramito, an individual cake 6 cm in diameter by 4 cm in height, wrapped 2, 6 and 20 cakes per bag. There are two varieties: the traditional white cake, and a dark cake. Ramito represented the remaining 13 per cent of the 1969 sales.

The perishability of the product necessarily limits the potential market of the cake to an area extending only 15 km around the Bogotá metropolitan area; thus the market is strictly local. There is a pilot sales agency in Pereira (a medium-size city 250 km from Bogotá). Distribution elsewhere is limited to nearby areas of the states of Boyacá, Tolima and Meta.

Distribution is by truck. Comercial Ramo Ltda. has 35 trucks, 25 of which deliver the product within the Bogotá metropolitan area. The truck driver and his assistant also serve as salesmen and are employed on a straight salary and commission basis. The Bogotá area is divided in 20 sales territories covered by the 25 trucks. The salesmen are responsible for adequate coverage and inventory renewal within their assigned territories, according to a sales quota per territory determined by the firm. From time to time, sales supervisors visit territories to check salesmen's performance. The firm has spotted some 17,000 retail sale points of which the company is covering around 15,000. Retail margins are not straight. They vary between approximately 10 and 20 per cent of the final price, depending on the item.

In general terms, the products are believed to have a high price elasticity. The basic marketing objective is to bring in a small profit per unit and to make profits through large sales.

The manager has a broad concept of the nature of his competitors: "any firm producing food products competing for the available minimum quantity of money to buy Ramo products". This seems to be a sound policy given the fact that some large and medium-size companies mainly producing wheat bread or biscuits also produce cakes, although in a smaller volume than Ramo. They are "late" competitors which have copied Ramo's success. Taking into account the fact that 62 per cent of sales are re presented by the traditional Ramo cake—mostly cut and sold by pieces by retailers—and slices, it seems logical that the firm should regard as a competitor any firm producing pastries, for example.

Promotion methods include taste panels, posters, special displays, and introductory offers at retail sale points. Newspapers, radio and television are used for advertising.

Costs in 1967 and 1969

Tables 12 to 14, showing sales, costs, salaries and productivity for 1967 and 1969, are based on data supplied by the firm.

As table 12 reveals, Productos Ramo SA shows an outstanding increase in total sales. Between 1967 and 1969 the increase was 47 per cent while the workers' cost-of-living index increased by 16.5 per cent. Most of the company's sales were made to low-income consumers. However, the increase in sales by products was not even. The traditional Ramo cake sales remained almost the same, although they still account for more than 60 per cent of total sales. On the other hand, sales of the individual cakes (Ramito) almost doubled, and those of the more fancy Gala cakes increased almost tenfold. The reason was perhaps that the market for cakes at low-income levels was

Table 12. Sales by products, Productos Ramo SA, 1967 and 1969

Product	1967 (thousands of pesos)	1969 (thousands of pesos)	1969 profit margin (%)
Ramo	14 960	15 500	11
Ramito	1 700	3 250	18
Gala	340	3 750	14
Mis Nueves	—	2 250	19
Hojuelas	—	250	36
Total	17 000	25 000	—

almost saturated, while there is ample room for developing the cake market at medium- and high-income levels.

Roughly, the cost of the raw materials represents 50 per cent of the total costs of the cake, although costs diminished slightly in 1969 (see table 13). Labour costs represented only 5.2 per cent and 4.5 per cent respectively in the two years considered. The main change was labour in the packaging operation (which is still performed manually). However, despite this fact, the relative importance of the labour force is small, and consequently it is not worth while to substitute capital for labour. The percentage cost in depreciation increased mainly as a result of the purchase of an oven. It is not possible to relate the main increase in depreciation to the diminishing labour cost, but only to the changes in packaging.

Other costs naturally increased, because of the enlargement of the installed capacity which is not being fully utilised. Merely between 1967 and 1969 the plant area more than doubled (from 2,021 m^2 to 4,429 m^2). The share distribution costs rose more than 20 per cent because of the broadening of the market, given the urban development of the city, plus the change in plant location from the downtown area to a site 20 km away.

The gross profit margin diminished, although the large increase in total sales assumes a large price elasticity of demand and the firm's aim to make profits through volume.

The direct labour costs per man-day (table 14) show the monetary salaries for the three main direct labour groups. In the fourth column, the salaries are adjusted upwards by means of the consumers' price index to give equivalent salaries in 1969 pesos. Without an increase in productivity, management would probably not have raised wages faster than prices. The rise in real wages could reflect a change in productivity. The growth rate for baking and packaging seems plausible. The mixing department was reorganised and mechanised and involves more complex factors.

Table 13. Cost distribution of Ramo cake production, Productos Ramo SA, 1967 and 1969

Costs	1967 (%)		1969 (%)	
Raw materials		51.9		49.8
Labour:				
mixing	0.73		0.72	
baking	0.71	5.2	0.69	4.5
packaging	3.76		3.09	
Depreciation:				
beaters	0.38		0.44	
ovens	1.82	2.6	2.64	3.6
packaging	0.40		0.52	
Others (fuels, power, water, parts, tools and maintenance)		3.4		4.4
Administration		9.5		9.8
Distribution		13.7		16.9
Total cost		86.3		89.0
Gross profit		13.7		11.0
Total		100.0		100.0

Table 14. Direct labour costs per man-day, Productos Ramo SA, 1967 and 1969

Process	Labour cost, 1967 (pesos)	Labour cost, 1969 (pesos)	1967 labour cost at 1969 prices (pesos)	Assumed change in productivity [1] (%)
Mixing	39.37	69.06	45.87	50.5
Baking	32.18	39.61	37.49	5.5
Packaging	30.10	36.87	35.07	5.1

[1] Assumed change in productivity $= \frac{\text{labour cost 1969} - \text{1967 labour cost at 1969 prices}}{\text{1967 labour cost at 1969 prices}} \times 100$

The value added in Ramo cake production was 7.2 million pesos in 1967 and 7.9 million pesos in 1969. This represents an increase of almost 10 per cent in the two-year period. In terms of Ramo sales, value added would account for 51 per cent and 50 per cent respectively. The value added per man engaged directly in Ramo cake production was 107,400 pesos in 1967 and 116,100 pesos in 1969. The daily production of Ramo cakes per man in direct labour was calculated at 238 cakes and 279 cakes respectively. This implies a 17 per cent increase in productivity per man employed directly in the process during the two-year period, or an 8.5 per cent annual increase.

If the increase in salaries attributable to rises in productivity were really 5 per cent for the baking and packaging groups, a comparison with the 17 per cent growth in physical productivity for the period under study would indicate that productivity has increased much faster than salaries. Productos Ramo SA has not matched productivity increases with wage rises, perhaps because it already paid higher wages than the average in the food industry.

PRODUCTION AND TECHNOLOGY

The production process is divided in four sections: mixing; dosimeters; baking and unmoulding; cooling and packaging. The mixing section belongs to the research instead of the production department.

The main ingredients are flour (imported grain milled in Colombia), eggs, sugar, shortening and animal fat, baking powder, emulsifiers, extracts and vegetable colours. Emulsifiers and vegetable colours are imported. Eggs and fats make up 85 per cent of the cost of raw materials. There is one basic formula for the production of the different cakes. Flavours are obtained by adding minor ingredients. A basic feature of Ramo products is that all cakes are simply the industrialisation of native recipes coming from older generations. Thus the technique of production has been developed in Colombia without any help from more advanced countries.

Since 1963 eggs have been broken in a machine with a two-man crew that inspects each one, breaks its shell from beneath and blows from above. The number of spoiled eggs has diminished substantially because of the industrial scale of operation of suppliers. The egg-breaking machine has three different speeds, being capable of breaking 4,000 to 6,000 eggs per hour. The daily production requires about 60,000 eggs. The firm also has a substitute machine for rush periods. The machines need only two men compared with the eight that were used before (the other six were reassigned at the same wages to a variety of tasks).

There are two types of mixer: one "planetarium" beater, and four small mixers. The planetarium beater took the place of five other small mixers; it has a capacity of 400 kg per hour and needs 14 persons. It mixes and adds air to the dough. The homogeneity of the dough is secured by means of a rotor that lets an even amount of oxygen into it, thus allowing change in density of the dough at will. It implies a continuous process. Given the importance of the traditional Ramo cake within the product line, this beater is used exclusively for its production. The other types of cake, which use a batch process, utilise the four small mixers that add air through beating. The amount of air added depends on time of operation and temperature.

In 1963 dough was still weighed manually. Today there are two types of dosimeter. The one used in the production of the traditional Ramo cake applies pneumatic pressure to conduct the dough from the mixing section to the oven. It automatically weighs the exact amount of dough per pan. Before letting the dough drop into pans, operators put a glassine paper in each so that the cake will not stick to the pan. Pans are set in the correct place manually, at the beginning of a belt that slowly leads them to the door of the oven. This operation requires 15 workers.

For non-traditional production, the mixed dough is brought from the mixing section in large containers placed on a hand-pushed barrow. Each container is then adjusted to a dosimeter that provides an equal amount of dough per pan. Since a lever must be pulled down by hand, the operation is slower than the pneumatic dosimeter. Nevertheless, two pans can be filled at a time. Pans are similarly covered with glassine papers and introduced to a second oven. It would be possible to supply dough with the pneumatic dosimeter simultaneously to both ovens.

Each oven is divided in five sections, each one of which has automatic temperature control. Burners are distributed in three sections and are not independent. The belt on which the pans pass through the oven has automatic speed regulation, set by a dial to between 5 and 90 minutes. Unlike the ultra-modern Belgian oven imported for biscuit production (see below), the two ovens for cakes are almost identical, although one is newer; the only difference is that the recent one is equipped for preheating, which implies a better quality of the baked product. The baking operation employs five people on each of the three ovens. In 1966, when there was only one oven, full capacity was used—two shifts for an average of 20 to 22 hours.

The entire production process depends on the capacity of the oven. The traditional Ramo cake spends 30 minutes in the oven. The production per hour is 2,100 cakes. Each cake enters the oven weighing 300 g and leaves it, already baked, weighing 270 g. The Gala and Mis Nueves cakes take 60 minutes, and are produced at a rate of 500 per hour. However (considering that they enter the oven weighing 1,000 g and that, baked, they weigh 920 g), in both cases production per hour in terms of weight is about the same. The bakers (really the only skilled workers in the plant) constantly check temperatures and times in the baking operation. At least four bakers are employed: one for each of the three ovens and one in a reserve position, which is rotated. The bakers were trained as in-plant apprentices under the supervision of higher-level plant technicians.

After coming out of the oven, cakes are unmoulded manually by 10 operators (5 per oven) and begin a cooling and hardening process that lasts 24 to 48 hours. The operation of arranging cakes to cool is performed by another 10 operators.

The packaging process begins the next day. Up to now, packaging has been performed by 40 operators. Most cakes are cut in triangular

(Ramo cake) and rectangular (Gala and Mis Nueves cakes) pieces. Although the cutting of rectangular cakes is done mechanically, it has been impossible to mechanise the cutting operation of the round cake.

Nevertheless, a major change in the cake-cutting system was introduced in 1963 with the help of trays having 12 guidesticks. Previously cutting was done by sight, which implied a slower operation and gross miscalculations in the size of each piece. This was particularly important as most cakes were (and still are) sold by the piece at retail shops. The contrast between the two innovations which the company was the first to introduce in Colombia is striking: the complex Belgian oven, and this simple cake-cutting aid. Technological progress covers a broad spectrum.

Cakes are sent on a metallic belt to packagers, who place them inside polyethylene or cellophane bags depending on the product, and finally other operators thermoseal each bag. Then the cakes are placed in metallic baskets and moved to the loading zone in barrows. Next day, in the afternoon, the trucks are loaded to begin deliveries to retailers early the following morning.

Thus cakes produced on Monday can be purchased by final consumers on Thursday. Cakes are produced from Monday to Friday. On Saturdays there is a general cleaning operation of the plant, and the cakes produced on the Friday are packaged. The cleaning operation is done manually, with hoses and air compressors, by the crew of one of the ovens.

Both ovens are currently working in the morning, but only one, in which the traditional cake is baked, works in the afternoon. On average the plant is working about 10.5 hours daily. Considering that the use of equipment at full capacity implies working in two shifts of about 10 hours each, the company actually is using 50 per cent of its installed capacity. Full capacity is used at seasonal peaks, such as Holy Week and Christmas.

Daily production is determined according to a day-to-day minimum programme of production that is the result of a careful study of the past year's sales revised to reflect the daily performance of each one of the products. Following such a plan, the necessary raw materials for a 15-day period are stocked at the warehouse, and the assignment of workers to the different tasks is planned. At the beginning of the year, the firm makes bids to secure suppliers and stable prices as well as the main raw materials, excluding wheat. Unmilled wheat is obtained through a quota from the Agricultural Marketing Institute (IDEMA). As a result, suppliers commit themselves regularly to provide specific quantities at prices determined according to the past year's average, plus or minus a given margin.

Quality control is secured by means of careful analyses at the different stages of the process: raw materials are sampled at the reception site; doughs in process are inspected four times a day; volume and standardisation analyses are performed at the laboratory twice a day; the baked product is inspected to determine crumb, texture, colour and bake. Every one of the Ramo products has a quality control seal put on by the cutter to identify the person who did the cutting operation. The general look of the cake is judged by the packer, who must discard every one having any imperfection. Finally, quality control of the product being sold is done by sampling. Through this step-by-step quality control, it has been possible to diminish substantially the number of defective cakes, which actually account for less than 0.5 per cent of total production. Every other day, salesmen change retailers' stocks to maintain freshness of the product. However, in good storage conditions and at an average temperature of 13°C, the product may last up to 22 days.

EMPLOYMENT AND WORKING CONDITIONS

The monthly payroll amounts to approximately 300,000 pesos, of which 181,800 pesos are paid to 197 workers and 119,000 pesos to 34 white-collar employees, including the general manager. Of each peso of output, only 4.5 cents represent direct labour costs.

With the exception of the bakers, the workers are unskilled. The simplicity of the tasks to be performed ensures that workers can easily be shifted from one job to another according to production requirements. The technicians are the food technologist and the mechanical engineering advisers, the directors of the research and the production departments, the analyst of the laboratory section of the research department, and the assistant supervisors and chief of the production department. The food technologist and the director and analyst of the research department work on the technical improvement of actual and prospective products. The remaining technicians work on maintenance, scheduling and control, and time and motion studies. All are professionals (predominantly engineers and chemists). The average white-collar worker earned 3,500 pesos per month, that is, 3.8 times as much as the monthly 922 pesos of the production worker.

The average wage per hour was 6.06 pesos (US$ 0.34) in 1969. The number of hours worked per week was about 50.

Productos Ramo SA has devised a salary curve based on efficiency. Seniority is a secondary issue to efficiency. A relative appraisal of each worker is made once a year to award a prize to the most efficient worker in the plant. The company's idea is to motivate workers to increase their productivity; thus the firm is willing to offer salaries above the industry average so that workers and employees can devote their full capabilities to solving the company's problems. Efficiency policy has not, in fact, led to a high turnover of personnel: about half the total number have been with the firm during five years or more. The number would have been larger if the firm had not grown so quickly in recent years.

However, turnover is high among the directors of the firm's six principal departments and among the technical advisers. Perhaps the manager is too authoritarian for these skilled individuals, whose scarcity raises pay differentials and lowers docility. Some considered that an "emotional one-man show" was intolerable.

In addition to legal fringe benefits (two extra monthly payments a year, social security services, and a family subsidy) the firm grants non-mandatory fringe benefits such as an additional monthly payment a year, family allowances, and special payments for birth, marriage or death in the family. In addition, the firm participates in the National Housing Institution (ICT) and plans to finance (jointly with the employee and ICT) the total value of the employee's home; its share is one-third.

Except for middle management, labour relations have been calm. An internal labour union, set up in 1967, has even helped to improve productivity in the plant. For example, the union had advised management to improve the workers' assignments in the plant to avoid waste. Every two years there is collective bargaining. It remains to be seen if this harmony will continue as the firm's growth slows, particularly with respect to employment as the labour-saving machines described below come in.

Although management has been interested in promoting employees, this has proved to be a difficult policy. Workers oppose the promotion of a fellow-worker to be their immediate supervisor, for example. It seems necessary to pick a worker from a different section for this, but workers even prefer to have superiors coming from outside the company. This attitude reflects the continued lack of the sort of union spirit that makes for ingrained respect for seniority and determination to keep all benefits within the group.

The firm provides special training courses to determine the feasibility of promoting workers. Moreover, it lets interested workers apply to the National Apprenticeship Service (SENA). Applicants can choose their own courses, but many are turned down because the courses selected are too little related to the work they perform.

Although discipline is very strict and little authority is delegated, and although a detailed concern for the personal welfare of subordinates is shown, the manager nevertheless disowns paternalism. He deplores that Colombian labour legislation is too paternalistic. Labour legislation is enacted with demagogic goals in mind, he says, and very little knowledge of workers' attitudes:

> There should be a managerial positive position towards workers, making them feel important members of the company. Most labour problems are the

result of wrong management attitudes. People are more important than machines; in fact, all capital should perform a social function towards the well-being of a group of people: basically the company's employees and suppliers. Friendliness is an important ingredient in obtaining positive labour relations.

On the other hand, he likes workers to know that he is continuously checking their performance, believing that productivity diminishes considerably when he is absent from the plant. In the future he expects workers to have a share in the firm's capital, but for the time being he feels that they are not prepared for this.

FUTURE TECHNOLOGICAL DEVELOPMENTS

Although the firm has advanced substantially since it began operations in 1953, management contemplates considerable technological improvements in the plant. The reasons for this are concerned much more with the goal of refining product quality than with making savings in a given operation. Considering the scarcity of savings and foreign exchange required for importing machinery to Colombia and the low level of wages, mere mechanisation is relatively unattractive.

Technicians at the firm are particularly concerned with the health problems resulting from the manual operations of unmoulding, cooling, cutting and packaging. However, some of these operations cannot be mechanised, at least in the short run. This is the case of cutting and unmoulding. On the other hand, cooling and packaging will be mechanised soon.

Although it would be easy to get appropriate machinery to unmould cakes mechanically, there are two factors preventing such a change: the quality of the fat being used to grease the pans, and the aluminium finish on the pans which is not smooth enough to allow mechanical operation.

As already mentioned, it has been very difficult to find a good machine to cut the round traditional Ramo cake. Apparently this shape of cake is quite uncommon in the industrial world. The actual production of 2,100 per hour and the need to make six cuts per cake call for about 200 cuts per minute. A machine to perform such work must be accurate and dependable, and the general manager has failed to find such equipment on world markets.

The firm is considering a refrigerated tunnel to simplify and speed up the hardening and cooling processes. Besides the advantages of diminishing the handling of cakes, the entire operation could be reduced to two hours instead of the 24 to 48 that it takes now. As a result, the life of the product could be extended by days and the stocks would be considerably reduced.

The machinery for packaging the round cakes has already been designed and ordered. It is feasible that the actual band being placed around each cake will be replaced by thin plastic plate of low cost. Instead of the present manual operation, the packaging machine will put each cake in a polyethylene paper and thermoseal it at both ends. This operation is expected not to yield large savings but to improve the appearance of the product. Displaced workers are to be transferred to the new cracker section where packaging remains manual.

Although the plant has been at its present location since 1968, the systems of receiving and handling raw materials, loading and unloading trucks and piling up full baskets for delivery are easily mechanised. Besides, the plant layout is ready for changes in the handling of raw materials to be used in daily production. The idea is to bring them up to a mezzanine by means of a lift, and organise the weighing operation there. Then the different types of ingredients already measured will fall from above to the beaters.

As already mentioned, the company is entering a slightly different market through the introduction of a new line of traditional biscuits. After a six-year period of development, the product is ready for its formal introduction on the market. The company has developed some 30 types of biscuit through the industrialisation of well-known native recipes that have, little by little, been disappearing as a result of the industrialisation of the food industry and the increasing number of women entering the labour force. At present the firm is selling some 2,000 kg per day of one particular type of biscuit.

The new biscuit line is completely different in operation from the cake line. The raw materials handling is the only part of the process that is similar. The firm imported a new Belgian oven with more advanced features: for instance, burners are independent in each of the four sections of the oven, the last burner being fed with gas recovered from the other three. Each burner can be adjusted to and maintained at an optimum temperature for its particular function. However, no comparison is feasible, because the oven is strictly for biscuits. Since November 1969, when the first biscuit was introduced, the firm has hired 33 new workers and shifted some of the senior supervisors from the cake line to the biscuit line.

* * *

What conclusion may we draw from this case? Perhaps it is that advanced technology can be introduced with ease in a country like Colombia for a commodity with an expanding market. Workers who are displaced in one part of a plant are easily reabsorbed in another. Wages can lag behind productivity as long as they are above the industry's or the national average.

No stimulus for developing a militant union spirit exists. Manpower problems arise, if anywhere, at the middle management level. Unlike the workers, technicians and middle management resent authoritarian superiors and will not be appeased even with high pay differentials. The result is a high rate of high-level turnover; but, at least in the case of Productos Ramo SA, not even this difficulty has hampered the rate of expansion.

ELECTRONIC DATA PROCESSING IN AN INDIAN TEXTILE FIRM

C. K. JOHRI
Shri Ram Centre for Industrial Relations, New Delhi

The purpose of this paper is to analyse the implications of introducing electronic data processing technology on manpower allocations within an old and well established firm, and the resulting industrial relations problems.[1] The paper is the outgrowth of a case study conducted by the author in the textile division of an 80-year-old multi-products corporation under Indian management with plants located mainly in northern India.

Data were collected through unstructured interviews with the key officials of the company. This method enabled the researcher to establish complete rapport with the respondents and provided the maximum opportunity for the latter to come out with information on the background to the introduction of electronic data processing and many aspects of the inner corporate life that could not possibly be obtained through formal questionnaires. The interviews, sometimes repeated, proceeded in an atmosphere of complete involvement of the researcher in the problems of the respondent and were consequently concerned not just with information gathering but also with analysis and interpretation in a climate of trust and mutual confidence. At the same time the researcher was advised not to interview union leaders or the employees concerned, presumably because to do so might compromise

[1] The term "automation" is used in respect of information processing technology rather than as a production process. It may be defined as "the replacement of human brains by versatile information processing machines". See Organisation for Economic Co-operation and Development (OECD), European Conference, Zurich, 1-4 Feb. 1966: *Manpower Aspects of Automation and Technical Change: Final Report* (Paris, 1966), p. 15 and the discussion, pp. 14-25. In the context of this case study the use of a computer is limited to the processing of information only as directed by the management and not with the transference of new data to the production process for determining operations. For automation in respect of servo-mechanisms, see Edward B. Shils: *Automation and Industrial Relations* (New York, Holt, Rinehart & Winston, 1963), Chs. 1-2; and *Automation and Technological Change: Hearings before the Subcommittee on Economic Stabilization of the Joint Committee on the Economic Report, 84th Congress* (Washington, DC, US Government Printing Office, 1955), pp. 10-27.

the position of the respondents. This advice was followed, and yet the researcher is convinced that it did not lead to the loss of significant information. The conviction stems partly from close familiarity with the industrial relations climate in the company and partly from the belief that the management had a good grasp of employees' reactions and the fact that respondents themselves had put them across in an unambiguous and straightforward manner. Accordingly the author, despite the use of subjective methods and one-sided interviews, has confidence in the results obtained.

One of the main conclusions of this study is that the primary effect of electronic data processing is on the structure of management and the style of decision making rather than on manpower or industrial relations.[1] The latter effects are secondary, and the problems engendered by electronic data processing can be overcome and resolved with relative ease, provided that the management is reasonably competent, that its communications with the employees are unclogged and clear, that its perception of their problems is generally accurate and that it is willing to pay the slight price of accommodation in making the needed adjustments in timing and emoluments. By comparison, the problems created for managements themselves are substantial, and the processes of adaptation, involving changes in procedures as well as power balance, are significant, even painful. In this paper, however, the principal focus is on the manpower and industrial relations aspects of adaptation rather than on the related problems of management. It should be noted that the apparently easy manpower adjustments could be serious, and might lead to explosive industrial relations, where management is deficient.[2]

SALIENT FEATURES OF THE COMPANY

The company started in the latter part of the nineteenth century as a cotton spinning mill. Later, particularly after the First World War, it added

[1] Its impact has been felt mainly in the development of new methods of scientific management involving application of mathematical techniques to the solution of business problems. Its effect upon the organisational structure is somewhat uncertain and has been projected to include increased centralisation, reorganisation of middle management, sharper demarcation of functions between the top and middle management, and so on. See Melvin Anshen: "Managerial Decisions", in The American Assembly: *Automation and Technological Change* (Englewood Cliffs, NJ, Prentice-Hall, 1962), pp. 66-83.

[2] As has been shown in the case of Life Insurance Corporation of India, Calcutta office. Here the question of the installation of a computer for servicing the insured clients, although accompanied by a management pledge of no retrenchment, led to considerable turmoil and loss of man-hours. The union contended that the high cost of computers with their known potential for displacing labour was bound to restrict future recruitment and promotion opportunities. The union leaders further held that automation defied the national goals of economic development and a socialist pattern of society. In the United States, by contrast, the progress of automation has not been hindered by much tension

weaving units and became one of the largest composite textile mills in India. At the time of Independence, the company had diversified into sugar, hydrogenated vegetable oils and heavy chemicals (such as caustic soda and bleaching powder) that were needed as inputs by the textile units. However, until the early 1960s the company's principal interest was in producing and marketing cotton textiles. During the 1960s the company, under the impact of the Second and the Third Five-Year Plans, branched out into wholly new products, such as PVC, rayon cord and fertilisers. In the process of diversification the textile units were dwarfed, creating a variety of stresses in the management structure. At the same time the unprecedented pace of expansion put enormous demands on the top management and led to its first tentative steps in the direction of modernisation, even experimentation. The stress of new challenges stimulated a desire among the managers to seek new methods, and a willingness to take criticism and to permit experiments with new ideas.

In part this willingness was made possible by high profits produced by an expanding domestic market that had been created under the aegis of economic planning. The years 1960-65 were marked by a steady increase in investment in both the public and the private sectors which, coupled with restrictions on imports and a growing population, created highly profitable conditions for domestic consumer goods industries. At the same time, the corporate income tax had been modified to provide ample incentives to take risks and to create new industries through a variety of concessions and loopholes. The national economic climate favoured rapid growth, and the Government urged the private sector to expand.

In this seductive environment, corporate management scrambled for growth. Failure to grasp new opportunities would have meant not just loss of potential profits, or the loss of esteem in the view of the authorities and the general public, but something even more significant—the loss of economic position relative to other corporations and business houses and the threat to existence that would necessarily follow. What it meant in practice was that the industrial licences issued in favour of a business house but not used by it might be withdrawn and reissued to another party. Thus, directly or indirectly, the Government encouraged all business houses having substantial managerial capability to seek industrial licences in accordance with the general priorities in the five-year plans; and, having obtained them, to work out collaboration agreements with foreign firms, to obtain the needed internal and external finance, and then, with as little delay as possible, to proceed with investment.

---

in industrial relations. See United States Department of Labor, Bureau of Labor Statistics: *Technological Trends in Major American Industries*, Bulletin No. 1474 (Washington, DC, Government Printing Office, Feb. 1966).

Inherent in this process was the considerable risk that some business houses might grow at above average rates and thereby establish pre-eminence in one product after another. Once they were in a field it was possible for them to block the entry of others. Indeed, by obtaining more than one licence one business house could smartly foreclose entry into new lines, sometimes even before the industries actually came into existence.[1] The danger it presented to already insecure business houses cannot be overestimated. The choice lay between rapid growth to maintain a relative position, or adjustment to a subordinate position in relation to faster-growing organisations. The problem was not of finance or marketing (these could be reduced to technical questions and solved) but principally of finding the requisite managerial personnel for the new enterprises.

The managing agency system

The company under study, like other enterprises then functioning in India (whether they were under Indian or European managements), was managed by a family-based managing agency house.[2] Under this arrangement the stockholders in an ordinary general meeting contract out all the rights and responsibilities of management to a firm usually registered as a private limited company. The latter would promote the new enterprise in the first place, subscribe or underwrite a substantial portion of the new issue and thus secure a sufficient block of votes for election as its managing agents. The managing agency is usually made up of the adult male members of a business family among whom the oldest, or in the event of a joint Hindu family, the most senior member of the lineage having the highest financial interests, is the chairman or the principal executive. On being elected as managing agents the adult members of the family are appointed as the key executives of the new enterprise. The managerial hierarchy of the firm essentially reproduces that in the family. In other words, the principal decision makers of the firm (be they known as chairmen, managing directors, directors or by

[1] See Government of India: *Report on Industrial Planning and Licensing Policy* (New Delhi, 1967). This report, as well as that of the Monopolies Inquiry Commission in 1965, brought out the role of industrial licensing in restricting the freedom of entry and the growth of concentration in the industrial sector.

[2] See Charles A. Myers: *Labor Problems in the Industrialization of India* (Cambridge, Mass., Harvard University Press, 1958). It was reported in 1967 that the total number of managing agents in India was 479, who managed 720 companies. The system has come under fire and was formally due for dissolution in 1970. However, the real switch to management through a functioning board of directors will probably take more time. See the Report of the Managing Inquiry Committee, 1966; and for an up-to-date brief account see S. C. Kuchhal: *The Industrial Economy of India*, 7th ed. (Chaitanya Publishing House, 1969), pp. 422-60.

some other title) are, mostly if not entirely, members of the family with the same order of deference, respect and even obedience as in a traditional Hindu family.

Occasionally, senior executives of the firm with a strong devotion and service to the family are elected members of the managing agency. This practice serves the double function of rewarding loyalty as well as creating the public image of secular and progressive management. This is also a device for augmenting the pool of top management talents when the family itself is not large, or not growing fast enough to undertake the responsibilities of an expanding business empire. All managers, and senior managers in particular, must adopt the values of service to the family and not just the organisation, for advancing their career interests. They must perceive family goals, develop a habit of loyalty to senior managing agents, integrate in their own minds the inseparable interests of the family and the organisation, and work for the advancement of both as essential for their own success. In so doing they seek to become junior members of the family and are often admitted to it.

The managing agency system thus creates managers in the mould of its senior members and perpetuates itself not merely through the family but even more powerfully through its managerial progenies and the self-contained organisational culture it creates. This is a mechanism of economising scarce resources in entrepreneurship, of building up large and expanding organisation networks around small nuclei, of overcoming the inevitable conflict between the entrepreneurial and managerial functions, and of motivating managers to achieve career goals through loyalty and identification.

This company had a board of directors composed of family members and outsiders, most of whom had an important financial interest in it. Since the 1930s the board has regularly accepted a worker into its membership. However, the board performs only the limited functions ascribed to it by law and has delegated most of the power to the managing agents. It will therefore be appropriate to say that, regardless of legal anomalies, the managing agents are the top management of the company.

The organisational structure is quite simple in form but complicated in practice. The company is divided into four divisions with directors as their chief executives. The chairman of the board of directors is the topmost executive and between him and the directors is the managing director. All the directors, as well as the managing director, are autonomous in their respective spheres and together comprise the managing agency.

Unlike the board of directors, which may meet as infrequently as once a year, the managing agency meets practically every month, and sometimes more often. The chairman of the company, as the most senior member of the managing agency, seems to enjoy overriding powers. However, all important decisions are based on family consensus. Below the directors are the general managers of the mills or plants, who supervise all functions (both line and staff) pertinent to the unit. The department heads report not only to the general manager but also to the director. The latter may call meetings of general managers, staff advisers and department heads, and collective decisions may be taken. These meetings may be formal or informal, of properly constituted committees or not, and may take place as often as necessary. However, once decisions are taken everyone knows his responsibility and the line relationship in execution is followed. The director may check on implementation through the committees' reports; criticism, discussion and decisions then follow. In this kind of organisational structure the director is the principal source of strength as well as of weakness; it is he who provides, or does not provide, new ideas and initiative. He sets the style of management, and everyone who is in direct contact with him has to face the choice, sooner or later, of either integrating with his style or leaving the company.

Leadership and internal change

If the director is bright, an innovator, a seeker of new ideas, able to set goals, and possesses the leadership qualities of recognising talent, of building teams of officers around clearly perceived tasks and of inspiring them to achieve them with maximum efficiency, the organisation as a whole will pulsate with energy and life. The consequences of deficiencies will be just as apparent. The director is the key resource; his personality, achievement orientation and style of leadership are absorbed by the managers and officers reporting to him and through them gradually permeate the organisation and determine its performance.

This means that in practice the same company may be differently led and managed in its different parts. The formal structures may be similar, but not the styles of management; and if the differences are pronounced, this may affect even the formal organisational structure. As can be seen, there is considerable potential for conflict within the larger organisations, and this may affect the working relationship between the family members as well. However, the principle of autonomy and the deeply ingrained habit of deference to age and seniority tend to mitigate stress and mollify tensions. Moreover, the probability of loss for the family that would result from splits and the ever-present threat from other business houses are usually sufficient to overcome tensions. It takes a great deal of tact, sagacity and wisdom on the part of the senior members of the house to hold the family and the organisation together.

A significant implication is that freedom and initiative accrue to the member who seeks them. Experiments involving substantial costs may be viewed with tolerance; failures may be pardoned and rationalised at the top as inevitable in the learning process. At the same time they provide the material basis for promoting a general belief that the organisation is progressive and that a measure of initiative is expected from all. The more entrepreneurial director may seize upon these values to implant the seeds of change as well as to accumulate power and influence. On the whole, however, the organisational climate in managing agency houses favours stability rather than change.

Early in 1960 the company did not have a specialised personnel function. Industrial relations were the direct responsibility of general managers who were, however, assisted by industrial relations and welfare officers placed low in the organisation's hierarchy. Two directors divided responsibility between textiles and chemicals. These had not yet been grouped into divisions with clearly demarcated jurisdiction areas and managerial responsibilities. The company had several able, experienced and loyal executives. Some of them drew high salaries, possibly among the highest in the country. It also had many officers, engineers and technicians, fully competent, well paid, and devoted to the company.

At this juncture, in April 1960, the eldest son (let us call him BRV) of the chairman of the company returned from the United States with a degree in business administration from one of the leading business schools in the Midwest and some practical training with a consulting firm in New England. He was in his early twenties and was immediately placed as a trainee in one of the textile mills. He worked hard, learned the trade and spent a great deal of time with the general manager. After six or seven months, he found himself totally dissatisfied with the state of affairs. Management as it existed was completely different from the theory of management that he had learned as a student. The need for change was urgent. He began by introducing double-entry book-keeping in one department after another, and had studies started on lapses in the blow room and on problems in the carding system and in the weaving section. The talent was already there, and methods were known; the problem was simply of introducing the change.

In November 1960 the silk mill of the company ran into rough waters. It had been sustaining losses, and the top management wanted to close it down but decided to give the young executive a chance to try out his ideas. BRV moved in and took decisive actions although he was nominally a trainee. He severed all connections with the central textiles marketing department (CTMD) and set up his own marketing strategy. Second, the general manager was dislodged and a new man, young like BRV and also back from the United States, was recruited. By December 1960 BRV had ceased to be a trainee and could take decisions on his own. He succeeded in silk.

By the end of 1960 other holders of degrees in business administration had been recruited and a new recruitment and training department had been set up with a young man in his twenties at its head. BRV had much influence in all this. In 1962 the discovery of malpractices in the mill in which BRV then worked led to several resignations and severances. BRV took over the textile mill as general manager and looked for replacements.

THE DECISION TO INSTALL A COMPUTER

By 1962 the internal climate as well as the composition of the company management had considerably changed. BRV himself organised several teams to study punchcards and accounting machines, and his experiments with linear programming in cotton blending showed that these could be tried on a continuing basis on a computer. It was felt that if a computer yielded a 10 per cent improvement in profitability it would justify the cost. Accordingly, BRV made a proposal to the committee of managing agents for approval to purchase an ICT computer. As it turned out, nobody took much interest in the request since it was considered too sophisticated. Nor, for that matter, did anyone object, and since BRV wanted it, the purchase was authorised.

As ICT was already in India, the company signed an agreement for model 1202 in October 1963 (a month later the order was changed to model 1300). At the same time an industrial engineer was appointed to be responsible for developing systems in addition to all industrial engineering activities in the textile mill. It was expected that the computer would be commissioned in December 1964 and so it was considered necessary to appoint six trainees in January 1964. However, due to procedural delays, the computer was not installed until April 1965. Although a core staff of four trainees had joined eight months earlier, it was some time before everyone became used to the new equipment and parallel runs began to yield successful results.

As has been noted, at the highest level there was little appreciation of what the computer could do. BRV, the young entrepreneur, generally knew what he wanted but was somewhat vague on the uses of computers. He was convinced, however, that the one they had could do a lot of things including finding solutions to linear programming problems. There is no doubt that he perceived it as a manifest symbol of modernisation which would gradually transform the philosophy and style of management. It is an instance of entrepreneurial decision under uncertainty. There was a general perception of gain, perhaps flowing from a fast calculating machine, savings made by the reduction of clerical posts, maybe the discovery of cotton blends that would reduce dependence upon imports and so on, but no carefully worked out plan. This is evident from the fact that the person appointed to take charge of the computer was an industrial engineer whose primary responsibility was to introduce more modern systems for maintaining records, checking stores, improving incentive wage payments in the weaving section and similar activities in which the company was woefully behind the times; this man considered something had to be done urgently to enable the management to face new challenges. Most managers were not consulted before the decision and some who knew about it accepted it as a directive from above. As one manager put it:

> The decision to have a computer stemmed from a latent desire to do something outstanding, i.e. research with an idea that with it one will be able to strike such substantial results that it will justify the cost. The concurrence was neither from below nor from above. "You do what you like"—this was the attitude. It was one man's decision.
>
> The image of being the first in the industry was dominant. BRV was more an experimentalist and probably wanted to project an image more of a researcher than of an industrialist.

Similarly, the general manager who replaced BRV in December 1963 explained in an interview that he was not involved in the decision-making process although the computer was going to be placed in his premises, charged to the profit and loss account of his unit and attend to his problems. As he recalls:

> It was a fast calculating machine which could do a lot of things. There was no planned view as to what the computer would do. The decision was: let the computer come, then we can do a lot of things; it will incidentally save a lot of clerical staff too. This subject was not discussed with clerks or their unions. The strategy of either installing or using the computer was not discussed with lower line managers.

The ICT local manager played the critical role of providing the expertise lacking in the company. He helped in preparing aptitude tests and had them sent to the United Kingdom for correction. He also helped in the final selection of trainees. He later worked with the industrial engineer on detailing systems and programming strategies. Furthermore, the ICT staff trained the trainees on the mill premises 10 months before the computer finally arrived. In other words, the need for new technology was perceived by the management while its operational and training requirements were taken care of by the supplier.

The reason for this gap between the perception of a need and provision of the technical requirements for fulfilling it resides in the degree of familiarity of management with modern office technology. The most modern office equipment the company knew was the Hollerith electrical accounting machine used in CTMD. This equipment was obviously useful but was not proving adequate for the tasks of CTMD. However, the marketing manager did not feel the need for a computer. When asked about the marketing problems, he replied: "There were none that the computer could help solve. We had a Hollerith machine.... The computer could help us through its greater sophistication." He added later: "The marketing department does not need the computer. Only a multiplier was needed. I was not consulted in taking the computer. I was only asked to make the computer useful and then we co-operated just as much as we could."

It is evident that at the managerial level there was a great deal of reluctance to have the computer at all. This caused delays and general dissatisfaction with its performance when it arrived. The industrial engineers who were given on-the-job training, and who later went for 11 weeks of training to the United Kingdom (in September 1965), complained of insufficient co-operation and even lack of appreciation of the reorganisation of tasks required by the computer. Meanwhile two senior technicians joined the computer centre but bungled the computerisation of office jobs. This led to increased mutual dissatisfaction between the computer centre and the user organisations.

To solve these problems, BRV, who had been promoted director in charge of the textiles division, appointed a Systems and Computer Group (SCG) in January 1966, signed an order for magnetic tape (which was expected to be delivered by October) and then left for a six-month tour abroad. This implied that SCG would have to sort out all the organisational snags without the support of the director. As it turned out, SCG wrote a detailed report on the matter and waited for the director to return. There was much uncertainty and insecurity over the role of the computer and why jobs were being computerised. This state of affairs continued through 1967 and 1968. Even the tape that was expected in October 1966 did not arrive. Consequently the computer centre and the managers closely associated with it came under severe attack. In the meantime, the textile division suffered serious market reverses and profits fell, lending weight to the criticism of an expensive and potentially disruptive machine. The young director found some of the most senior managers ranged against his style of management. This reached a point where he could not count upon the support of his family members, not even his father, and he grew desperate and at one stage considered resigning from the company. It appeared early in 1968 that the computer might be dumped. However, luckily for the company, this did not happen. BRV won through against his opponents and secured greater freedom and power for himself. These were used for effecting internal transfers and promotions, consolidating the head office of the textiles division and obtaining greater information control through the computer. Since the ICT 1300 was found inadequate for the job and the ICT organisation in India unequal to the task of meeting the requirements of a growing computer centre, it was decided to change to IBM. In September 1968 an IBM 1401 with tape was installed in the reconditioned and enlarged premises of the computer centre. The former industrial engineer was now designated manager of the computer centre and had more staff and greater confidence, felt more secure and enjoyed a closer rapport with the user organisations. By the middle of 1969 most of the organisational snags had been solved and the computer was fully utilised; indeed, the company had to rent time on more advanced IBM computers available in the area.

So the computer finally won, and made decisive changes in the organisational practices and systems. It should be noted, however, that in all the battles that raged no one left the organisation and no effort was made to accelerate someone's retirement. It is part of the paternal culture of the management that once someone has integrated himself into the common network of family and the organisation, he stays in regardless of defeats. A basic loyalty transcends all other considerations. One may be angered at a decision, even feel outraged at its arbitrariness or at not being consulted, and in turn may criticise or cause delays; but this is an internal matter and the stress by itself would not justify severance from the family/company relationship. If a general manager does not get along well with the director in charge of his division, he will be left alone or may start reporting to the chairman instead. Alternately, he may leave one unit and take on another, and virtually run it just as he pleases. Given the fundamentals of loyalty and ability to produce profits and to keep the enterprise working at an acceptable level of efficiency, there is much room for diversity in management practices, including even the structural forms of organisations. This is unity in diversity, as is shown by the fact that although the computer has come to stay, not all organisations that could use it were doing so in late 1969. There was little pressure upon them to conform.

MANPOWER ADJUSTMENTS AND INDUSTRIAL RELATIONS

Managerial and supervisory levels

The decision to install a computer led to two complementary decisions:

— to appoint an industrial engineer with potential for handling the managerial tasks of: (*a*) recruitment and training junior staff; (*b*) co-ordinating with the supplier of equipment at the technical level; (*c*) advising management on the economics of computers and all related decisions; (*d*) developing systems of book-keeping, data reporting, collection of needed information and reporting, etc., so as to render the computer useful to the production and marketing organisations on the one hand and to higher management on the other; (*e*) building up team work and good public relations (this was a key appointment);

— to appoint two senior technicians in the officer grade for supervising the machine and the staff; these joined the computer centre three months after the installation of the ICT 1300.

Subordinate manpower

The operative staff of the computer centre consisted of two groups: trainees recruited directly, and staff transferred from CTMD. These were punchcard and other machine operators, checkers, computists and others employed on the Hollerith electric accounting machine and rendered surplus following the transference of work from CTMD to the computer centre.

The trainees were recruited in batches. The first group of four trainees joined the company in July 1964. Earlier, in December 1963, 24 candidates were given aptitude tests in Calcutta, Bombay, Madras and Delhi. Again in September 1965 six systems trainees were recruited. The former were trained by the ICT systems adviser. The latter were given on-the-job training by the supervising staff. The systems trainees were used for training and helping accountants, clerks and others in developing suitable forms and procedures for data storage and reporting. Formerly, information pertaining to production, efficiency, wages and marketing was kept in leatherbound registers from which it was difficult to retrieve relevant data quickly. The change from the old to the new meant irritation, misunderstandings, mistakes and complaints. The computer centre deployed three men to clear up these troubles in July 1966. In October of the same year, when the computer centre took on the computerisation of the costing system for the oldest mill, serious troubles arose because of reporting errors and incomplete systems data. Consequently two trainees were deployed to streamline operations.

Although it was decided as early as December 1965 to transfer the work of CTMD to the computer centre, it actually took much more time than had been anticipated. The marketing manager made a comprehensive request for reports and analyses from the computer centre on the grounds that it ought to do jobs that the Hollerith could not. There was no disagreement on the importance of the reports asked for; however, it was another matter to develop systems to render them feasible. After much correspondence and negotiation and many meetings, the work of CTMD was finally transferred to the computer centre in January 1967. The terms pertaining to costs and staff were fixed by the marketing manager and agreed to by the computer centre. Under this agreement, all key punch operators (whether needed or surplus) were transferred to the centre in the following months. In May 1967 the Hollerith was closed down. The clerks were mostly absorbed in the same organisation, a few being transferred to other units. By August 1967 the industrial engineering department had been separated not only functionally but also physically, involving transference of staff.

It may be noted that the transfer of CTMD manpower to the computer centre coincided with the increase in machine utilisation above one shift. The growth in workload is indicated by the fact that expenditures on cards and stationery increased from Rs 1,500 [1] in the September-December 1966 quarter to Rs 8,000 in the first quarter of 1967 and Rs 11,000 in the second quarter of the same year—that is, by more than seven times. It was therefore possible to find work for most people acquired by the computer centre.

Between July 1967 and January 1969 the computer centre recruited 97 persons in all categories and acquired 25 men from CTMD. Of the former, a few left and some went with the industrial engineering department. At the beginning of 1969 the computer centre had a complement of 46 employees, comprising the computer centre manager, 11 officers, 2 supervisors, 2 control clerks, 6 computer operators and 24 key punch operators, verifiers, computists and tabulator operators. The total expenditure on the computer centre was about Rs 1 million, a negligible fraction of the total outlay of the textile division which ran into millions of rupees. At the peak of controversy in 1966 the cost of running the computer centre was about Rs 600,000. It had loomed large, however, because of falling profits, which in 1966 were a mere 40 per cent of the 1964 level; everyone looked around for a convenient scapegoat to cut costs. But taken by itself the cost of running the computer centre was well within the means of the company.

The job descriptions and necessary educational qualifications described in table 15 indicate no scarcity of candidates for the jobs. Indeed, under Indian conditions, the position is quite the opposite. The minimum educational qualification, the Higher Secondary Diploma, could be met easily. The supply was virtually perfectly elastic at the prevailing salary level. The supply of experienced operators was limited but nevertheless sufficient to meet the scant demand. In no case was much training required. The most

[1] US$ 1.00 = Rs 7.5.

Table 15. Jobs, job descriptions and qualifications in the computer centre of an Indian textile firm

Job designation	Job description	Educational level	Experience required
Clerk	To carry out ordinary clerical work such as postings, totalling and simple calculation	Higher secondary	
Key punch operator	To punch or verify data from handwritten sheets in prescribed card designs. Should be able to handle any type of card designs and both numerical and alphabetical punching. To work with hands as well as automatic machines	Higher secondary	
Computist	To carry out simple addition, multiplication, division, subtraction, etc., with hand-operated comptometers	Higher secondary	
Sorter operator	To sort punchcards on the standard sorter	Higher secondary	One or two years of key punching
Tabulator operator	To wire control panel of the tabulator and do tabulating jobs	Higher secondary	One or two years as sorter operator
Computer operator	To operate the computer according to instructions conveyed through a job scheduling register	Graduate	Orientation programme by IBM

skilled job (computer operator) needed a training period not exceeding one month. Consequently, the company had no difficulty finding the manpower it needed.

Indeed, through the transfer of staff from CTMD and inexact estimates of manpower needed by the computer centre—due partly to inexperience and partly to excessive reliance upon the advice of computer company executives—it ended up with a moderate surplus. If the company changes to more advanced models, the surplus will probably grow rather than diminish. It is probable that the company expected some turnover, but this has been negligible and can be easily explained. First, the company has a policy of paying a salary differential above the prevailing industry-cum-region rates. This policy enables it to attract above-average applicants, assuring better selection. However, once a person has been taken, it is not the policy to ease out those found to be surplus. Second, the computer business in India is still in its early stages. As the number of computers multiplies, the demand for experienced hands will grow and may generate some turnover. Until then, employees with a couple of years of work behind them in fairly specific kinds of job would be most reluctant to leave. By insisting on their departure the company might create an industrial conflict with wider significance and unpredictable consequences.

In view of the uncertainties through which the computer centre had passed, which inevitably engendered a feeling of insecurity therein, the manager probably considered it prudent to ignore the problem and to concentrate on more pressing issues. Since top management did not insist on tightly defined manpower charts anyway, he did not view it an urgent matter either. Apparently everyone understood that with the transfer of 25 men from CTMD, some loose manning would have to be reckoned with as the inevitable cost of securing the concurrence of the marketing manager on the one hand and more effective utilisation of the computer on the other.

Indirect manpower effects

At the subordinate level the introduction of the computer led to considerable changes in job content and work relationships, though in very few cases did it either create new jobs or render existing ones surplus. Let us take the former kind of change first.

To begin with, the impact of the computer was felt by only those organisations that wanted jobs computerised. Since the computer centre was organisationally part of the textile division, its presence was noted but not felt in other divisions of the company. Even within the textile division, the computer was pointedly ignored by at least two units, both of which were located at a distance of more than 100 miles. One of the general managers was positively hostile to the very idea of tampering with the established flow of information and the control points in the organisation and so, although he could not dislodge the computer, he succeeded in immunising himself and his unit. At least one other general manager successfully fought it off, first, by letting the computer centre do the production reports and the payroll and then finding too many faults in them (most of which had crept in due to malfunctioning of new systems), and second, by arguing that instead of effecting economies, the computer centre had saddled the unit with the added cost of maintaining two systems, which in turn was causing delays and embarrassing the management. So he carried on his work without the computer, and while this meant temporary disruption of work for clerks in the payroll and the production departments and some anxiety to the senior production personnel concerned, the old system nevertheless reasserted itself. This did not mean there was no change, however; experience with the computer had exposed the weaknesses of the traditional systems and these had to be modified in consultation with the industrial engineering department. But the unit did free itself from the tyranny of remote control and the loss of power and initiative that the computer represented.

The general manager of this unit did not argue against the computer but instead took the line that "a thorough systems study should have been made and a master plan prepared". When asked about the likelihood of retrenchment he replied:

> The issue of retrenchment has not been raised. But, as there was a possibility of reduction of clerical staff, the strength of permanent staff was frozen. This affected the work in departments. When a clerk was required, a temporary hand was taken who had no commitment to his work, and so the work suffered. However, the computer has not been put to any use that will reduce clerical strength. The cost of doing the same work by the computer will be much higher than the saving effected by the reduction of staff. Hence no reduction was made. When the computer can give new information that highlights the shortcomings of a department it is useful.

What he wanted the computer to do was to make sales forecasts and give him the ideal product mix. In the absence of these it was more a nuisance than a help.

In the oldest production unit the computer changed not only the job content but also the number of jobs. For the clerks maintaining production records on looms, it meant changes in the method of work, involving greater discipline and adherence to time schedules as well as irritation and displeasure arising out of reporting errors. Furthermore, the clerks faced the problem of explaining to the workers (who usually approached them individually) how they had performed on given looms, both in relation to their own past record as well as to the performance of other workers. Previously a clerk could open the pages of a leatherbound register and satisfy the workers. But now he could not.

The situation was made worse in that the detection of errors on the printed page sometimes resulted in short payments to workers on incentive wages and there was a risk of loss of mutual confidence; and also by the fact that at first neither clerks nor officers could make head nor tail of the printed statement. The production superintendent of the mill explained the situation as follows:

> There were a lot of difficulties. The immediate problems that we had to face were the errors and discrepancies in the wage statements that were being prepared on the computer. Work was transferred to the computer, a large amount of work done by the clerks was stopped and the number of clerks was reduced.
>
> There were a lot of mistakes made by clerks in filling up the tables. These were genuine mistakes, which formerly were overlooked. But when the computer statements started coming in there was some problem of satisfying workers about their accuracy.

He solved the problem by asking the complaining worker to fill in a form and having it passed on to the weaving clerk, who had to re-do the calculations manually. This took about three to four days. He also pointed out that, as a result of mistakes, some clerks had to be reprimanded although no one was punished.

The clerical strength had been reduced from 62 to 42 (i.e. by almost 33 per cent). However, no one had to leave the organisation because vacancies arose elsewhere. Since the clerks were company employees, and transferability at no loss of income was a condition of employment, the company faced no problem in relocating them.

It is significant, however, that the mill did not eventually discard the old system of maintaining registers by looms, sorts, shifts, sheds, sections, and so on. After much dissatisfaction with the services rendered by the computer and complaints of high cost, the old system was revived. At the same time the original clerical strength was not restored. This meant that the clerks who remained shouldered heavier workloads at no extra pay while the company saved a little on the salary bill of transferred clerks and on supervisory time. Thus, indirectly, the computer helped in cutting feather-bedding and in raising efficiency in the weaving section.

Net manpower effects

Managerial and supervisory staff. The computer caused a direct accretion of strength and also had indirect effects on the uses of managerial and supervisory time. In the computer centre itself the company hired one manager and 11 officers. As the computer centre took on the research and analysis of production, inventories and marketing problems, it found this complement insufficient for its needs. In 1969 the computer centre was seeking more expertise at higher levels of training and achievement to be deployed on original research. This pointed to further employment of highly paid manpower.

In the user organisations the computer brought about substantial differences in the work of shift officers and the first-line supervisors or *mistries*. If the computer had rendered faultless service some of these functionaries might have become redundant. However, this did not happen and well before they needed to put up organised resistance, in one mill the computer services were withdrawn and in another the old system was revived and functioned parallel to the computerised systems.

At the higher level the work of general managers has changed markedly. There is no longer any excuse for groping in the dark. It is now possible to streamline cotton purchases; the probable results of blending different varieties of cottons and of cotton and synthetic fibres on the final products can be ascertained with reasonable accuracy; the level of efficiency in the loomsheds is known and is not a matter of conjecture (it is even possible to know which looms yield profits and which do not); and the management has a better idea of the inventory position.

Actually, since production planning was done by CTMD on an ad hoc basis, the management knew only approximately how many qualities and varieties the company produced and had far less idea of their exact inventory position and profitability. Everyone was surprised when in 1969, for the first time, the computer centre disclosed the exact number. It was so large that the management could not consciously even have considered producing them, to say nothing of estimating their costs and profitability. Hitherto the management had been geared to just two tasks: to keep production going, and to show satisfactory profits at the end of the year. However, the steady profit squeeze starting from 1965 (made worse by the successive droughts of 1966 and 1967), coupled with the sharp curtailment in imports of Egyptian cotton, forced upon the management the need for cost and profit planning for the first time. This meant strengthening the accounts and research and development functions, as well as integrating the industrial engineering section more closely with the management.

The personnel department was also developed, again for the first time, and gradually found acceptance in the organisation. This meant not only more personnel—by itself a very small change in relation to the total—but also a far greater change in the style of work and in the allocation of management time. The computer did not cause these changes, but rather accelerated the pace once the direction was set. Once installed, the computer centre could not but move the organisation in the direction of modernisation and create new awareness of the problems that had existed all along but had been ignored. It provided new information, reduced general problems into specific questions and furnished the relevant data needed for their solution.

Office staff. The introduction of the computer led to a direct increase in employment of 10 persons in the computer centre (an insignificant factor in relation to total employment, which ran to five figures) and to a freeze in the strength of clerks. Therefore it adversely affected the potential employment of white-collar workers and perhaps somewhat reduced the promotion opportunities of clerks as a whole. It should be noted, however, that promotion opportunities for clerks had always been few and it was normally expected that most of them would not go beyond their fixed time scale of pay. The pay, allowances and time scale could, no doubt, be raised from time to time, as a result of either collective bargaining or indirect pressure leading to an industry-cum-region survey and a unilateral decision by the management, but that was about all that they could expect.

In fact the company did not bother about saving on clerical costs. First, the managers, by and large, were opposed to it; they did not perceive it as the legitimate consequence of a computer. They wanted to have more rewarding results and more effective guidance from the computer centre, rather than to waste their time on clerk relocations and the resulting grievances so close to themselves. Second, facts spoke eloquently against this policy. As the chief accountant of the textiles division pointed out, the oldest mill was paying the computer centre about Rs 500,000 per year for rendering the statements on production, loom efficiency and payroll, and this could not possibly be offset by a reduction of clerks. Even if 20 clerks were taken off the payroll, the net saving would be about Rs 85,000 per year, whereas a 1 per cent improvement in the loom efficiency would mean a saving of Rs 100,000. So the computer centre would justify its cost if it raised weaving efficiency by 4 to 5 per cent. There was really no point in relating savings on clerks to the computer centre and the gains that might accrue from it.

Industrial relations

Mainly due to the management protection afforded to the clerks the computer centre made virtually no impact on industrial relations. To begin with, the attitude of clerks, and one they shared with their officers, was to "wait and see". Later there was some insecurity, but this feeling too they shared with their officers. Their supervisors and middle managers took up their cause, argued for them with senior managers, and protected their interests in a manner they could not possibly have done themselves. Since no one lost his job, no major grievances arose. The only group that came close to creating a potential industrial relations dispute was the one transferred from CTMD to the computer centre. Throughout the latter part of 1967 and 1968 there was a simmering, even growing discontent; and they had a case.

First, at CTMD the working conditions were much better than at the computer centre. Whereas at the former, clerks were expected to put in 6½ hours per day and 39 hours a week, at the computer centre these were 7 hours a day and 42 hours per week. So transfer meant longer hours of work. Second, at CTMD the workflow was not as arduous as at the computer centre. The latter required far greater attentiveness, care and adherence to schedules than at CTMD. This was, however, an underlying cause of discontent rather than an articulated grievance. Third, there were certain fringe benefits, such as entertainment pay and compensating leave for overtime work. The entertainment pay was small and not a regular feature but was paid whenever clerks stayed longer to finish their work. There were no such fringe benefits at the computer centre, and no one to care for the transferred people. Lastly, transfer to the computer centre added to the commuting time and expenses of most of the affected persons. They wanted compensation for these losses.

For about a year the management paid little heed to these grievances. But late in 1968 it began to appear that the grievances might be formally taken up by the clerks' union. It also seemed that the clerks were winning the support of other employees of the computer centre. In fact, with the installation of the IBM 1401, some staff had become surplus. Though no one had been informed formally, they could perceive the danger in the diminished flow of cards and the leisure time some employees had on their hands. A bonus dispute affecting all the workers and clerks was also brewing and a strike appeared a distinct possibility. In view of this, management decided to compensate the transferred clerks for the increase in their working hours, by giving them two graded increments. It appears that this gesture by the management was taken as a fair compromise and the threat of dispute was withdrawn.

There was no other instance of a grievance that assumed the proportions of an industrial relations problem. This is largely because the number affected was small in relation to the size of the company, because no one actually lost his job or took a cut in pay, and because the clerks enjoyed the support of their seniors, with the result that no cleavage between the ranks occurred and the lines of communication remained open throughout.

CONCLUSIONS

It is always hazardous to generalise on the basis of a particular case for the economy as a whole. In practice there is no "representative firm" whose behaviour may be taken as characteristic of the industry. While most firms, particularly the large ones, do no doubt share some common features, in a way each one of them is unique. Each firm is conditioned by its peculiar background, the philosophy of management, and the force of customs and habits prevalent among the managers and employees, and will consequently respond differently to a given stimulus. This will be the more pronounced the more isolated is the stimulus itself and the smaller the number of people it affects.

As shown in the case above, the computer remained a subject of curiosity for a majority of workers and clerks until it began to make mistakes in printed wage statements. Even these mistakes remained in the realm of bona fide errors of a remote staff function which the line organisation was, however, prepared to remedy. Things might have been different if a computer with a larger capacity had been installed in the first instance, if the management had insisted on making clerical savings pay for part of the cost, if the anomalies discovered in the incentive pay system had led to attempted revisions to the disadvantage of workers, and if a change in the marketing strategy resulting from computer research had led to an overhaul of the marketing organisation involving turnover of personnel. But these are speculative "ifs" and are not relevant to a case study even though they raise interesting probabilities. In a case study the focus of research is not on what might have happened, but on what actually happened and on the conclusions to be drawn therefrom.

In spite of these problems of generalising, four conclusions with policy implications nevertheless warrant mention:

1. The use of electronic data processing places severe pressure on traditional managers to modernise their systems. Its impact is so sharp that, in organisations unaccustomed to change, the introduction of such technology may be possible only through an authoritarian management decision, although its acceptance in a progressive organisation would be facilitated by group discussion.

2. From an economic standpoint, the cost savings for the firm must be sought in some area other than a saving on the number of clerks employed. Given the relative cost of wages for such employees and the cost of using computers, even a large displacement of office personnel can hardly cover the expense of the new technology. It can be expected therefore that computer utilisation will focus, even more than in industrialised countries, on its potential in areas other than labour costs.

3. The case study shows that the problem of training personnel for computer use was relatively minor. There is difficulty, however, in generalising from this experience to draw conclusions about the training problems which might arise in countries without the same educational resources as India.

4. With respect to potential problems in industrial relations, these do not pose insuperable obstacles to the use of modern technology, particularly in an expanding firm where management is cognisant of the human issues involved in introducing change.

COMPUTERS IN EAST PAKISTAN

A. FAROUK
Dacca University

At the time of Independence in 1947, East Pakistan had no non-traditional manufacturing industry except for a few cotton textile mills. By 1963-64 the total number of factories (as registered under the Factories Act) was 2,010 and the average daily employment in these factories was around 256,000. Nearly half of this employment was in textiles. (By contrast, the civilian labour force numbers about 17 million.)

Although the economy is still primitive and modern technology still in its infancy, there are nevertheless examples of automation in some sectors of the economy. This study will review some of the manpower effects of the introduction of computers in the Dacca area.

Twenty-five years ago, most office and accounting work was done manually. After Independence, ordinary calculators and other mechanical aids came into use. In government offices in East Pakistan, Facit calculators were first used in 1950. Three years later, a set of Powers Samas tabulation machines was installed in the Bureau of Statistics. Some business houses also obtained calculating machines, with the European business firms generally leading the way. For instance, Messrs. McNeal and Kilburn had a set of tabulating machines in use in Chittagong as early as 1955. The Institute of Statistics at Dacca University and the Adamjee Jute Mills have installed tabulating machines more recently. These are isolated cases, however, and most of the business houses continue to do their accounting manually, in the traditional manner.

In the past few years, however, a new development has taken place: the introduction of electronic computers. In 1964 the Government's Atomic Energy Centre in Dacca brought an IBM 1620 computer to the province. This was the first computer in Pakistan, and it was used almost exclusively for the centre's research. In 1967 the United Bank installed an ICT 1901 computer in Dacca and allowed other firms to use it. The Habib Bank installed an IBM 1401 computer in Dacca in January 1968. An IBM 1401 computer was also installed in July 1970 by the Adamjees, who control and manage firms in several industries along with the country's three largest jute mills.

Finally, in December 1969, the United Bank set up an ICT 1902 computer for their branch and clients in Chittagong. Altogether, the United Bank had 28 computer clients in June 1970, and the Habib Bank had nine. The work consisted mainly of preparing bills and payrolls and of inventory control. Both banks acquired these clients within a year of computer installation. Both had also installed a computer in Karachi a year before doing so in Dacca. Since the installation of the United Bank and the Habib Bank computers, the Atomic Energy Centre has also sold computer time to business houses.

To obtain information regarding computer use, the executive heads in the computer bureaux were interviewed through a questionnaire including the following items: date of installation; purpose of installation; structure of management; method of staff recruitment; kind of job being handled and intended; if working for others, names of organisations now using the computer; structure and size of the organisation; staff salary scales; problems faced in computer use; future plans.

The executive heads of the organisations buying computer service (but not themselves owning computers) were then interviewed through a questionnaire including the following items: structure of management; nature of job given over to the computer; reasons for the decision; number and grades of workers employed in the organisation; staff training; salary and methods of recruitment of the persons previously doing the job; policy and problems regarding the unemployed; problems of machine redundancy, if any; new recruitment; salary and training of staff needed for new jobs created by the computer; advantages and disadvantages of computer use; workers' reactions to computer use. To supplement this information, a few employees from each organisation were interviewed to ascertain their reactions to the computer and to verify the information provided by the executive head.

Interviews could not be held in one organisation, the Chittagong Water and Sewerage Authority (which has used the computer service since December 1967), because Chittagong is 200 miles from Dacca. Information for this enterprise was partly provided by the United Bank computer bureau. As interviews were held in similar organisations in Dacca and Narayanganj, the loss of information is not considered serious. Each interview took about two hours, and in two cases some follow-up information was obtained through a second interview. The author acknowledges his indebtedness to the respondents from these organisations, but for whose frank responses to his questions this study would not have been possible.

USES OF COMPUTERS

Until mid-1968 the United Bank computer operated for only about 10 hours a day, using only half its capacity. The jobs fell into three categories:

— payroll preparation, accounting and computations for the bank itself (in Dacca and Narayanganj, the bank now has 28 branches, in addition to the head office);

— the once-for-all type of tabulation and computation job that is occasionally required (for instance, in December 1967, the bank made a compilation of paddy acreage and yield for each crop in East Pakistan at the request of the Government of East Pakistan);

— work performed on a continuous basis—this type of work expanded rapidly and had a direct and considerable impact on employment; it consisted of preparing payrolls for a semi-governmental organisation sponsoring and owning the majority of the large industries in East Pakistan, for the largest cigarette manufacturers in Pakistan with two factories in East Pakistan, for the newly established Chittagong Steel Mill, for Dacca University, and for others.

The uses of the Habib Bank computer fall into the same three categories:

— in the first category, the bank prepares its own payrolls and also uses the computer to identify "prize bond" winners among its clients. (Prize bonds are a kind of loan floated by the Government of Pakistan. Instead of interest being earned, prizes are won through a lottery. The bank registers all prize bond holders among its clients. If any of the bonds they hold secures a prize, the bank locates it on the computer and informs the client concerned. This is an example of a service that the bank can offer to its customers only because it has the computer. There is no doubt that this device has saved thousands of bond holders from having to search for their numbers in the very long list of prize winners issued once every three months.) The bank also collects applications for land allotments of the Dacca Improvement Trust under the Uttara township scheme and keeps track of them;

— in the second category, in April 1968 the bank completed a census of civil employees of the Government of East Pakistan for the Eastern Railways for the year 1964/65;

— in the third category, the bank took up the work of bill preparation of Rajshahi Electric Supply in August 1968, and has negotiated a number of jobs for the payrolls, inventory control, stores accounting, etc., of industrial and commercial clients.

Some firms do not have computers in East Pakistan but are using their Karachi computer for their total business. The Pakistan International Airlines are an example.

Although the experience of computer use in East Pakistan is not yet substantial, some cases of the impact of the computer, particularly with regard to earnings, employment and training of manpower, can be examined.

THE MANPOWER IMPACT

The impact of computers on labour in East Pakistan can be examined from the angles of training, quality, levels of earning and the quantitative aspects of employment, and the impact on management. We shall first examine the training aspect.

Clerical work was always done manually. A typical clerk in East Pakistan is the lowest rank of white-collar employee. Although many clerks are arts graduates, in most cases the degree of educational attainment is a high school leaving certificate. At present, there is an abundant supply of people with this type of education. During the British rule there was very little opportunity for technical and scientific education. There were no engineering or medical colleges and very few technical schools. On the other hand, there were nearly 100 arts colleges and about 2,000 high schools. There was very little vocational, scientific or commercial education; and even in 1966 there were only 27 professional educational institutions. In 1965-66 the number of students enrolled in the general degree colleges in the province was about 100,000, whereas the number enrolled in the technical institutes and polytechnic institutions was only 4,000.

A medical or an engineering graduate usually starts at a monthly remuneration of Rs 400 [1] after 10 years of schooling plus 6 years of study after high school. An arts graduate, on the other hand, with 10 years of schooling plus 4 years of later education, could only expect a monthly salary between Rs 100 and Rs 150.

The traditional system

A matriculate clerk now starts at about Rs 110 per month. A typical clerk receives in-service training in the place of employment, and his salary rises to a maximum of about Rs 200 at the age of 55. In most cases his job involves copying, maintaining registers, manual computation involving addition

[1] US$ 1.00 = Rs 4.76.

and subtraction, and incidental office correspondence. His work is countersigned and checked by an officer whose salary is at least double his own.

The Dacca Water and Sewerage Authority (WASA) was this kind of office before its billing section went over to the computer in 1968. It had the following structure: one chief assessment officer earning Rs 630-1,050 per month; below him, three assessment officers with Rs 450-800 per month (these positions went only to the most senior graduates); below them, five assessors (these were graduates with a pay scale of Rs 225-550 per month); and below them were 30 lower division clerks who were matriculates and paid on the scale of Rs 110-240 per month. (A factory worker earns between Rs 80 and Rs 150 per month, paid weekly in most cases.)

This kind of organisation is, therefore, bottom-heavy. As the bulk of the work reaches the lower division clerk first, this man determines the speed and standard of the work in the whole office. He has no incentive to work better and more quickly, as he earns only a little more than a factory worker (although for the latter the only educational requirement is literacy). In addition, a lower division clerk has the problem of maintaining a certain standard of living because he is a white-collar man living in expensive urban areas. The clerks therefore form a group of employees living without much hope, with poor food and in perpetual poverty. The position is little better in the offices of non-Pakistani-managed business organisations.

A few companies (some with European management) pay a little more to clerks. In the Pakistan Tobacco Company, a typical clerk starts at Rs 300, and although he is only a matriculate he can move up to section head clerk, when his salary would rise to Rs 684 per month. But these are exceptions, and the salary of a clerk and the absence of incentive cited in the case of Dacca WASA is typical of many of the autonomous, semi-governmental and private business houses in East Pakistan.

Office structure in computer bureaux

Compared with this system, the office structure in a computer bureau is far less bottom-heavy. The employees are much better paid and more skilled, as a result of their formal training and apprenticeship. For instance, in the United Bank computer bureau, the organisation is as follows: the top man is the EDP (electronic data processing) manager. Under him there are three systems analysts, ten programmers and an operations manager. The EDP manager and the three systems analysts are chartered accountants trained on the job elsewhere in computer operations. The systems analysts receive about Rs 2,000 per month and the EDP manager is still more highly paid (although on contract service). The ten programmers are mathematics and statistics graduates and start at Rs 500 per month, with the prospect of doubling their income in a few years. The punchers and operators who work below them are matriculates, earning about Rs 350 per month. There are a number of documentation clerks who are matriculates earning Rs 300 per month on the average. Their supervisor, a graduate, is paid Rs 450 per month. Regarding training, it was reported that the first batch of staff recruited for the United Bank in August 1967 were mostly fresh from the schools and the colleges, except for four people on the administration side who were transferred from other branches within the bank. There is in-service training for the staff; an incumbent is paid a regular salary only after he qualifies in the training, which goes on for about three months. It was stated that the reason why most of the computer staff were recruited directly was that young people were more adaptable. Furthermore, they were available in the (not very large) numbers needed by the computer and at the attractive salary that was offered.

Information concerning the structure and level of the pay scale in the Habib Bank computer bureau has been compared with that of the United Bank. Although the pattern and the initial qualification vary slightly in the former organisation (for instance, some of their systems analysts are masters of business administration and not chartered accountants), the salary of staff and the structure of the organisation is about the same. Regarding training, those operating the computers in the two Dacca banks reaped the benefits from the experience gained earlier in Karachi. Some trained staff for punching, etc., came from the Atomic Energy Centre computer bureau at Dacca, which had installed a computer in 1964.

The following generalisations are thus possible:

- the jobs of clerks who are replaced by the computer staff are comparatively numerous, less well paid, more monotonous, and far less attractive;
- the jobs of different categories in the computer unit require a higher level of intelligence and intensive training;
- there is very little possibility of absorbing the men displaced by the computer by offering them employment on the computer itself, since the number of people required is far less and the qualifications are far higher than those required of a typical matriculate clerk working manually in an office;
- interviews with the lowest cadre employees of both types of organisation indicate that the majority of clerks in the offices (before a computer is installed) are relatively more frustrated, ill-clothed and, on the basis of looks at least, ill-fed, with very little job satisfaction. On the other

hand, the employees in the computer section are better dressed, more satisfied with their income and perhaps better fed. The employees working on the computer, however, said that while at work, they had to be very alert and they were supervised more strictly. Compared with this, the clerks of traditional systems feel less tension in their work and are much less closely supervised;

— in terms of management, the traditional pattern is inferior to the management in a computer bureau. This is because, first, in a traditional office there is hardly an objective criterion to check the quality of work, and second, it is not possible to reward a good worker by giving him better remuneration.[1] Although the office clerks do not earn very much more than factory workers, they rarely have strong workers' unions. In some offices, although there is no legal bar, unions have been discouraged by the management.[2] Management in a computer bureau is more streamlined and scientific, compared with that in the traditional offices. The computer is run by very few people, paid relatively better. As yet they do not think of forming a union.

IMPLICATIONS FOR THE FIRM

What benefits derive from the use of electronic computers? This question was asked of all the enterprises using any of the computers owned and hired out by the two banks. The answers varied, but the important points can be summarised as follows.

Most of the enterprises had no previous idea of the gains that they would have. The Pakistan Tobacco Company is an exception because their Karachi office had already been using the computer for two years, for payroll preparation. The initial temptation to use the computer in Dacca came when the two banks offered services on a "free trial" basis. In general, there had been very little use of machines earlier. Only three of the clients were using some kind of mechanical device: the Pakistan Tobacco Company was using the Facit and Underwood Add-Listing machines; the EPIDC had the

[1] During the interview, an exceptionally intelligent but frustrated lower division clerk remarked: "In our office I do not work to my capacity, because if a man is known to be a good hand, he has no reward, as there are no posts above the rank; but his application is not forwarded elsewhere for a better job, he is not granted leave on the plea that he is an essential hand, and the superior officer allots more work to him because he is prompt."

[2] It is reported that in the East Pakistan Industrial Development Corporation (EPIDC), for instance, the head office staff is not allowed to form a union. In Dacca University there is an association of clerical and subordinate employees, but its registration was cancelled a few years back for irregular submission of reports. It has not yet been revived.

Kenzil accounting machine, which meant only partial help; and the Dacca University accounts office had been using two Facit calculating machines borrowed two years earlier from the statistics department and retained since then. The machines were inadequate for these organisations, considering their total volume of work. Very few redundancy problems would have arisen if part of the work had been handed over to the computer. For the first month, during the parallel run, the same work was done by the computer and by the existing staff, to compare the quality and to experiment. Subsequently, however, most of the clients were convinced of the advantages of computerisation and have since been thinking of transferring additional work to the computer.

It appears that the computer is not cutting down employment very greatly, at least for the present. The supervisory staff has remained intact in all the offices. Only at the clerical level has there been some labour saving (excluding the cost of computer use).

The EPIDC, for instance, formerly used accounting machines on double-shifts. Payroll preparation alone kept these machines busy in two shifts for seven days in the month. Now the computer does the work in about an hour. This has brought relief to the machine room. Previously the section preparing payrolls consisted of seven persons: one accountant, one assistant accountant and five clerks. After computerisation, three clerks out of five were transferred to other branches that were understaffed. In terms of the number of workers, the cut was about 40 per cent; in terms of pay, it was about 20 per cent because only the lowest-paid clerks were released. The staff could not be reduced beyond this, because although payroll preparation was mechanised, the change statements or correction statements had to be sent to the computer every month. The correspondence work also remains unaffected.

In the case of the Pakistan Tobacco Company, four clerks were supervised by a section head clerk and an officer above him. After computerisation two clerks were retained in the section and the remaining two transferred to other sections. The two clerks retained have to prepare the information for the computer, check the cards, note changes, resignations, increments, promotions, etc. The experience of the Pakistan Tobacco Company was thus comparable to that of the EPIDC.

The experience of the Dacca WASA

To illustrate costs and gains associated with computer use in East Pakistan, the case of the Dacca WASA can be described. This organisation was set up in 1963 to modernise and improve water and sewerage services in the developing city of Dacca. WASA took over the existing works, assets and liabilities from the Dacca Municipal Corporation and organised itself as a separate and autonomous body. WASA expanded rapidly, and by 1969 the total number of accounts handled came to 35,000 (including about 3,800 government accounts). In 1959 there had been only 8,000.

The revenue division of WASA makes assessments and collects bills. Salaries, as reported above, ranged from Rs 1,050 per month for the chief assessment officer to Rs 110 for the lowest lower division clerk. The latter were the lowest level of staff working in the office—preparing bills, keeping files of accounts, maintaining various ledger cards and registers, and drafting correspondence with clients.

In 1969 the organisation of this section was changed slightly and the staff redesignated. The chief assessment officer is now called the chief revenue officer, the assessment officers are now called revenue officers and their number has been increased to four. The assessors are now called revenue supervisors. The lower division clerks still go by the old name.

WASA sends quarterly bills to 35,000 clients or account holders. Until April 1968 the entire job of preparing bills was done without the use of any mechanical aids. At that time the job of preparing the bills for only one experimental area (Dhanmondi) was turned over to the United Bank computer bureau on payment of the cost of stationery alone, which was nominal. By September 1969 the entire work of billing went over to the computer, except for the 3,800 government accounts.

Before computer use, each lower division clerk was given 700 accounts. His job involved the entry of new connections in the ledger cards, maintenance of the index register and the street-wise register, entry of newly assessed rates, recording of changes of address, maintenance of the billing register, preparation of the fortnightly billing statements and quarterly instalment bills, and so on. As the new connections increased very rapidly, WASA became a bottom-heavy organisation, was unable to despatch the bills in time, had inadequate information about collections and had little time to press defaulters. Writing bills alone took two months out of each quarter.

The bank, using its computer, now collects the data from WASA, prepares the bills and submits a quarterly statement of billing and a monthly statement of collections to WASA. Account holders can pay their bills through any branch of the bank in the city. In the future the bank may also undertake the work of issuing notices to defaulters.

Although savings were not very evident in 1968, they can be calculated for 1969. There was no monetary saving at the supervisory level. At the clerical level, 21 clerks have been released from this division and have been transferred to other branches. Lower division clerks on the average earn about

Rs 200 a month (including fringe benefits like house rent subsidies). The saving in a year, therefore, is about Rs 50,000. The computer cost is Rs 4,000. Thus, the monetary saving is Rs 46,000. Regarding the indirect gains, WASA feels that it now has better information and more time to devote to defaults. Bills can now be sent in about a week's time. Although at the initial stage mistakes were frequent, the computer is now more accurate. Accounting is more speedy and much work has already been brought up to date. The firm is spared the bother of sanctioning new posts with every increase of work. As yet the firm has not followed up the possible advantages that might accrue from the provision by the computer of more cross-tables for intensive analysis; when WASA has more financial analysts and experts the study of these might lead to better decision making.

A disadvantage, however, is that WASA is now dependent on the bank for financial records which it formerly held itself. The risk here is that for every item of basic information it has to look to the bank, an outside organisation. In a few cases WASA has lost money when an account holder has left and WASA could not bill him directly in time (although it knew that the account holder was leaving). Nor did the bank send him the bill, because the account holder did not communicate with the bank for that purpose.

The total cost of this revenue division is about Rs 125,000; the over-all financial savings in the division are about 37 per cent, and the savings in terms of employment are above 50 per cent. It is possible, however, that the bank may later increase the service charge for its computer, which would eat away a part of the financial gain (the savings on employment, of course, remaining unchanged).

Dacca University

Dacca University gave the job of preparing the monthly payrolls of its 1,700 employees to the United Bank computer in April 1968. They did not gain much immediate advantage from using the computer and have stated that it upset their entire system of operation.

The Dacca University accounts office disburses not only the salary of the staff but also a large volume of subsidiary bills on examination remunerations, etc., to the teachers. Therefore, only a part of the work has been taken over by the computer. The registers have to be kept according to the old system, where all payments to the staff in addition to salary have to be recorded for deduction of income tax, which is determined by the total income of a person and deducted at source. As the university is a non-profit-making institution, there is hardly any incentive for accounting analysis of the data. Moreover, since items like "club dues", "cost of medicine" and so on are variable deductions, new information has to be given to the computer every month, and this has meant no change of employment in the accounts office. There are now seven persons in the bill section of the office: one senior assistant (salary about Rs 275 per month), three upper division clerks (salary about Rs 200 per month), and three lower division clerks (salary about Rs 110 per month). It seems that the university will have to change its method of accounting and also change the present system of programming if it is to continue using the computer. The university did not pay anything to the bank for the use of the computer service except Rs 250 per month to cover the cost of paper.

Advantages and disadvantages

Advantages. The advantages, on the evidence of those who are now using the computer, are as follows:

— office work, which implied monotonous copying by the clerical staff, not necessarily with speed, accuracy and neatness, can now be speeded up (delays in the disbursement of salaries and in the collection of bills are disliked by all business organisations);

— formerly there was virtually no analysis of the wages and collection data, because tabulation was expensive. Now that the basic data are already on the computer tape, all cross-classification and group totals can be had in minutes. This makes possible better budgeting, better analysis and forecasting of future requirements, better analysis of arrears, quicker location of bad debts—in short, better over-all management of the business;

— control of corruption in certain cases where speed provides a fast check on accuracy. For example, in public utilities a speedy recording of payments prevents the granting of discounts to late payers.

To give some concrete examples: one organisation drawing up its payrolls by computer found that it could make departmental surveys, provident fund calculations, budgeting and credit advice to banks much more quickly; another organisation having bills prepared by the computer now knew where the areas of bad payment were and which classes of consumers were important, and in some cases it was able, for the first time, to collect huge unpaid bills, simply because the staff now had more time to look into the files. The computer serves as eyes for the businessman.

Recently there has been increased public criticism of inefficiency in semi-governmental and government-owned business institutions. Fortunately this criticism has coincided with the availability of a computer service in East

Pakistan. The use of a computer has brought relief in terms of managerial work, where there is a shortage of trained manpower. With very little marketing effort, the banks are therefore finding enough clients to keep their two computers reasonably occupied.[1]

Disadvantages. Although a considerable volume of work from various kinds of enterprise has already been shifted to the computer, the disadvantages of its use are no less conspicuous. Many of the present users of the computer are not sure if it will be really economic. It appears that the economies arising from computer use have been greatest in well-managed business organisations.

The owners of the computers, however, feel certain that their use will be economic for all clients. At first they did not charge many of the clients because they thought that the actual cost of operation would be determined more definitely when the banks knew the extent of the service performed after the volume of work increased considerably. Originally the United Bank made a small charge of Rs 250 for each month's paper. Although the banks indicated their reluctance to quote any final rate for use of the computer, one of their clients whose West Pakistan office had used the service for the previous two years disclosed that they had to pay between Rs 400 and Rs 500 for one hour's use.

One hour's use of the computer, however, if properly programmed, can release at least five lower division clerks for other work. At this level, the wage saving is equivalent to the computer charge. The possibility of obtaining different types of cross-table from the same data is a net gain for the client. This advantage, however, can be taken only by an enterprise that is analysis-minded, and requires qualified and trained business researchers. These increase the cost to the enterprise but also ensure higher returns through better decisions. As one of the computer owners stated:

> There will be an increase in the requirements of higher-level staff for our clients, as each one of them would need a computer division in their enterprise. We are, however, running courses to train their staff in the numerous uses of the computer, in the preparation of their budgets, in planning and measuring their fluctuations and in the comparison of various segments that is now possible from data analysis and interpretation.

Effects on manpower as seen by employers

On the question of unemployment following the introduction of computers in East Pakistan, the experience of most of the respondents in the interviews

[1] In West Pakistan the Government is examining the possibility of transferring the centuries-old treasury and accounting system to the computer, and a recent report indicates that this is quite possible: see *Report on the Treasury and Accounting System of the Government of West Pakistan* (Karachi, United Bank, 1967).

indicated beyond doubt that there would now be fewer jobs for the clerical staff, although clerical jobs would not be abolished altogether. Some clerks would be needed for corrections, for adjusting the work for the computer, and for preparing correspondence regarding defaults, etc. As one of the respondents pointed out, "Certain jobs would not be given to the computer, because of their confidential nature—like bank balances, bank advice, etc., even if such work was time-consuming manually."

But it was also clear from the evidence of all the respondents that the cutting-down of clerical jobs would not lead to any unemployment of the existing workforce. This is paradoxically true, because at the moment all the clients of the computer, as well as the banks themselves, are expanding their activities very rapidly. Against this, the computer is taking away only a small part of their total office work. Even in Dacca University, which is not a business institution, the expansion of work has been so great that surplus clerks can immediately be used in other sections. It is understood that between 1955 and 1968 the clerical strength in the office of Dacca University increased by 100 per cent. In the case of commercial enterprises, the expansion is at least equal. This is not to say that the computer will have no impact on employment. Many organisations have stopped recruitment at the clerical grades because of automation and rationalisation measures. The employers in East Pakistan now find frequently that, if two or three posts of clerks are advertised, several hundred candidates with the required qualification will apply. On the other hand, for technical jobs, the employers very often do not get as many candidates as there are jobs. There is an acute shortage of manpower in East Pakistan in all sectors of skilled employment, except for jobs requiring only matriculates or arts graduates. Most of the respondents admitted that switching over to the computer requires trained people who are in short supply at the moment. The banks who are hiring out the computers are, however, conscious of this problem and are helping their clients to train their existing staff.

There are also some peculiar technical problems for the computer owners themselves. Finding spare parts and servicing the computers in cases of major breakdown are still difficult, because the total number of computers in the city is so few. Another factor peculiar to the East Pakistani owners of computers is the frequent fluctuation in electric voltage in the city. It was pointed out by the owners of the computers that voltage variation and frequent electricity cuts spoil their work and damage their machines. One of them is already thinking of setting up a stand-by generator. This need will make computer operation relatively more costly in East Pakistan.

CONCLUSION

The foregoing sections have examined the economic and social background of East Pakistan and noted how automation is increasingly entering the sphere of office accounting through the electronic computer. This change has taken place in spite of the primitive technological level in many other spheres of the economy. The examination of the effects on manpower of automation in office accounting has led to the conclusion that, in spite of the numerous advantages of computerisation of accounting work, automation is having the undesirable effect of reducing employment prospects for a particular class in the labour force that, by an historical accident, is in abundant supply at present. It seems obvious, however, that automation in this sphere of the economy is inevitable and will increase in magnitude in the near future.

The following are a few of the more important problems resulting from the introduction of computers in East Pakistan:

1. In view of the disproportionately large supply of the matriculate and arts-educated labour force (as compared with technically skilled personnel), computerisation is going to create unemployment in this section of the labour force in the immediate future.

2. Computers upset an economic class structure that is already unsatisfactory. The income groups in urban society at the moment may be roughly divided into four categories: group 1, with a monthly income of Rs 100 or less ("workers"); group 2, with a monthly income between Rs 101 and Rs 250 ("lower middle class"); group 3, with a monthly income between Rs 251 and Rs 500 ("middle class"); and group 4, with a monthly income above Rs 501 ("well-to-do"). In urban East Pakistan men in the lowest two groups are already too numerous. As we go to the higher income groups, numbers decrease sharply.[1] If clerical employment is reduced, the inevitable result will be to push men down from the second group to the first group. This will widen the differences between economic classes in our society, where the rich are very few and the poor are already too numerous. In the interests of social peace, it would be better if we could increase the middle class and reduce the proportion of the lowest income group. An unemployed but educated middle class is a danger to any society.

3. In the Western world the computer originated from the need for more objective decisions based on intricate analysis of routine data. Com-

[1] According to A. Farouk and M. Safiullah: *Retailing of Consumer Goods in East Pakistan* (Dacca University, Bureau of Economic Research, 1965), the income distribution of a random sample of urban dwellers in Dacca in 1965 was as follows: up to Rs 100 per month, 38.80 per cent; Rs 101 to Rs 250 per month, 36.32 per cent; Rs 251 to Rs 500 per month, 16.41 per cent; above Rs 501 per month, 8.47 per cent.

puter operation is uneconomic for developing countries unless the equipment is properly and fully used. Technically competent manpower and trained business administrators are in extremely short supply in East Pakistan. This being the case, the loss in the level of employment is not likely to be compensated by better decision making in the short run. One can expect that computers will do only what the clerks were doing before.

* * *

Whether desired or not, computers have appeared in East Pakistan. They should be used for the more intricate type of work, which some respondents mentioned only as part of their future plans. Probably no additional machines should be introduced unless the existing ones are used to their full capacity.

The introduction of automation through computers in East Pakistan is therefore creating technical problems of skill shortages, higher cost of operation due to the undeveloped nature of some of the services and inadequate volume, and some use of automation in institutions that really do not need such highly specialised techniques. It also results in fewer employment opportunities for a section of the labour force which comes primarily from the lower middle class of the society. These are serious problems which are likely to have considerable repercussions on the society and economy.

ELECTRONIC DATA PROCESSING BY ETHIOPIAN AIRLINES

SEYOUM SELASSIE
Haile Sellassie I University, Addis Ababa

Ethiopian economists, social scientists and planners hold that close to 90 per cent of the population of Ethiopia lives in the rural areas and is engaged in agriculture and allied activities. There are even those who think that this estimate is rather conservative because most people living in small towns make their livelihood from agricultural occupations and therefore should not be regarded as belonging to the urban population.

Industry is not widespread, being concentrated in the Addis Ababa, Asmara, Assab, Diredawa and Winji-Nazareth areas. These areas are major forces of attraction for the rural population in the adjacent hinterland. And yet the employment opportunities these industries offer are extremely limited. This small demand for labour is illustrated by the fact that during the fiscal year 1966-67, when the population of Ethiopia was 24 million, the total number of people employed in modern industries was 58,000 [1], excluding 900 foreign employees. In 1966-67 there were 395 more or less modern industrial establishments in the country. A division of employment by establishments shows that on the average each establishment employs 150 persons.

Close to 70,000 people are employed in governmental services, making this area larger than industry, without including the armed forces. A considerable number of people work either as daily labourers or for fixed periods of time in the private building industry and on public works, but there are no records to show the actual numbers. Another source of employment, primarily in urban areas, is domestic service. Again the number of people engaged in such activities is not known. There are also people engaged in shoemaking, basketry, weaving, pottery and other cottage crafts, but once more no information is available about the number of people engaged in such activities.

[1] Ethiopian Central Statistics Office: *Statistical Abstract, 1966.*

Currently there are several governmental, semi-governmental and non-governmental organisations that use IBM computers, including the Bank of Ethiopia, the Imperial Board of Telecommunications, the Ethiopian Electric Light and Power Authority, the Imperial Highway Authority, the United Nations Economic Commission for Africa and the Central Statistics Office.

It is against this background that the effects of automation on manpower should be examined. The subject of this case study is Ethiopian Airlines, one of the most progressive commercial undertakings with a workforce of approximately 3,000 people.

GROWTH OF ETHIOPIAN AIRLINES

Ethiopian Airlines was established as a company in 1945 and flight operations started in 1946. The airline was to help break up regional isolation within the country and to help Ethiopia's export trade. In those days Ethiopia had a limited number of roads connecting the provinces to the capital. Only one railway line ran between Addis Ababa and the port of Djibouti in what was then French Somaliland.[1] Ethiopia had no other access to the sea. The airline system was a part of an over-all national development strategy.

During its early days, the airline directed its efforts primarily towards domestic needs. It started by connecting major towns with Addis Ababa and, in addition to this domestic service, began flights to regional cities like Cairo, Aden and Bombay.

Over the years, especially in the late 1950s, domestic services expanded and flights to Europe became regular. As Ethiopia developed, trade with Europe expanded and tourist potentialities came to be recognised. It therefore became necessary to strengthen the airline fleet. Around 1958, two DC6B aircraft were added to raise the quality of service on the European route, which was then extended from Athens to Frankfurt. Currently Ethiopian Airlines connects Addis Ababa with Nairobi, Entebbe, Dar es Salaam and Lusaka in East Africa; with Khartoum; with Monrovia, Lagos and Accra in West Africa; and with Cairo and Beirut in the Middle East. It also flies to Karachi, Delhi, Rome and Madrid. Freight services have also increased considerably, and cargo flights operate between Addis Ababa and European cities.

In 1946 Ethiopian Airlines carried only an estimated 8,000 passengers, as against 110,000 in 1961 and 208,000 in 1967. Passenger miles increased

[1] Now the French Territory of the Afars and the Issas.

from 3 million in 1946 to 71 million in 1961 and 177 million in 1967. Freight ton miles increased from 300,000 in 1946 to 2.8 million in 1961 and 7.4 million in 1967. Stockholders' equity increased from Eth$ 528,000 [1] in 1946 to Eth$ 15.9 million in 1961 and Eth$ 29.5 million in 1967.

Ethiopian Airlines was established in collaboration with the Trans-World Airlines system. Managerial and technical personnel were provided by TWA. At the same time, plans were made for a gradual phase-out of foreign personnel by giving training to Ethiopian nationals. The main outside source for the recruitment of Ethiopian nationals as pilots was the Imperial Ethiopian Air Force. Almost a dozen fully-fledged international jet pilots are Ethiopian nationals. A number of Ethiopians have also assumed responsibility in high-level management. Many others are in the secretarial services.

Introduction of electronic data processing

This growth of the airline called for a more efficient and rapid method of processing information in order to provide prompt services both to customers and to shareholders. As the number of employees of the company also grew, accurate knowledge of the performance records of all employees became imperative.

As Ethiopian Airlines officials put it, for the airline to be competitive in a world-wide system, quick and accurate decisions have to be made. This would be possible only if there were a quick and accurate way of marshalling information. Therefore it became necessary for the company to install computers to process financial and other information in the most effective manner. Bookings, however, are not yet made by computer.

The computer was installed in the Treasury Department towards the end of 1964. The following six types of IBM machines are currently operating:

— 029, for card punching—this serves as a punching and automatic interpretation unit;

— 1447 (Consul), a programming device which considerably cut programming time and after its installation led to a much higher degree of efficiency;

— 1311 (disc system), a machine with dual functions—it transfers information from the punchcard to the information storage disc or vice versa;

— 1442, a card repunching machine—its function is to repunch cards on the basis of information fed into it after programming;

[1] US$ 1.00 = Eth$ 2.50.

— 1441, the final processing unit—the computer itself. Once all the data are put in order, it takes approximately one hour to produce the information required by departmental executives;

— 083, a manual sorter, used when the volume of sorting is too small to warrant the use of machine time.

These machines are used for finance purposes, attendance records and operating statistics. Before the computer was installed all this information was handled manually. It took a long time for executives to get all the facts needed for decisions.

A few female employees carry out the routine work of punching, repunching and typing, under the supervision of a well-trained and experienced programmer. Many of the employees who were previously working in accounts were also trained to operate the computers.

All lower-level staff in computer operations were given training in the country by IBM Ethiopia. People selected for supervisory positions were trained abroad.

The introduction of the 029 machine can illustrate the circumstances under which a more advanced technology replaced an earlier system (in this case, the manual punching of cards). It is interesting that neither the possibility of labour savings nor that of raising volume justified replacement within the first five years of installation.

Six operators and one supervisor carried out the manual punching in 1964, and the same number operated the 029 in 1969. Meanwhile the number of key depressions rose from 6,000 to 12,000 per hour. Company officials felt that the seven workers might still have been able to handle the larger volume of work manually. Information was not released about wages before or after the change, but presumably there was no decrease; hence wage savings could not have motivated the shift. The real advantage of the 029 was that it provided automatic interpretation of the data. If a separate interpretation unit had been rented, its monthly cost, oddly enough, would have been more than that of the 029: US$ 100 compared with US$ 65. Since the manual equipment was not without value, the monthly savings have probably been over US$ 35.

MANPOWER QUESTIONS AND THE COMPUTER

When plans were developed for the installation of the computer, Ethiopian Airlines was aware that automation might result in displacement of workers. Since it is one of the biggest commercial enterprises in the country, it needed to reconcile two apparently irreconcilable factors: as an organisation with international operations, its need to be concerned with efficiency; and as a major employer, its need to avoid reducing already scarce employment opportunities.

When the decision was taken to install the computer, the company was careful to seek a minimum of displacement and a maximum of financial advantage. That is why, apparently, the Treasury Department was selected first.

The company was convinced that the installation of the computer would be beneficial to the employees. There would be greater opportunities for promotion and salary increments for those retrained for operating the computer. Moreover, in the days when the computer was installed, urban unemployment was not as pressing a problem as it is now, because the tempo of migration was slower.

Company officials state that there were no manpower problems at all when the computers were installed. The computer was a totally new piece of equipment and the employees might not have been aware of the possible consequences of the innovation. Ethiopian employees were not familiar with the intricacies and implications of automation, and the installation of a computer was probably regarded by the employees as just another addition in office equipment. In other words, the possible consequences were not readily apparent to them. The fact that the innovation was on a small scale may have avoided employee concern over job security. In fact, the prospect of being selected for reassignment may well have given rise to new hopes. Those were also the days when the idea of unionism and collective security of workers had not taken root in the minds of wage earners. Perhaps the situation would have been different if the installation of the computer had taken place now, because wage earners in general, and those employed in the private sector in particular, are more conscious of their rights than they were several years ago. Many workers now suspect hidden motives behind managerial actions.

By the time that unionism had gained in strength and a new consciousness of the collective power of the workers had developed, the operation of the computer was already a *fait accompli*. They could do nothing about it. An example of this new kind of consciousness can be found in the strike of workers on a sugar estate against a proposal to mechanise the harvesting of sugar-cane. These are, of course, workers engaged in a totally different industry, and the introduction may have been less tactful. Yet the incident gives a clue as to how the employees of Ethiopian Airlines might have reacted if the introduction of the computer had taken place five years later.

The absence of employee reaction could also be attributed to management's skilful handling of the situation. Employees were adequately prepared for the introduction of the machine; fears of negative consequences were dispelled. In Ethiopia the most important concern of workers is security, while wage considerations are almost equally important. In this case the employees were reassured on both questions.

Effects on staff advancement

Ethiopian Airlines officials believe that the installation of the computer in the Treasury Department has, in fact, opened up new vistas for the employees of that department. Working with the computer entails new responsibilities and requires a new pattern of behaviour on the part of the operator. It requires a more sophisticated level of skill meticulousness, and a new concept of time.

Traditionally, the notion of time did not carry as much meaning as it has for people in technologically advanced societies. Therefore, responsibilities in computer operation represents a major shift in outlook.

Those whose aptitudes and inclinations gave sufficient indication of their likelihood of success in the new situation were given opportunities for training and assignment in computer operations and data processing. Others were transferred from other departments and given training for programming and supervisory posts within the Treasury Department and the data processing unit. These transfers were needed because too few workers in the Treasury Department had enough aptitude and ability for computer operation. Ethiopian Airlines has its own training facilities.

Advanced technology was introduced in two stages: first, calculating machines in 1962, followed by the computer in 1964. Before 1962 practically all accounting processes were done manually with such mechanical aids as simple calculators, tabulators, and so on. There was very little data processing in those days. The 1440 computer started operations towards the end of 1964 and the beginning of 1965. In 1961, 77 people were employed in the Treasury Department, none of whom were engaged in data processing as this had not yet been introduced. By 1962, with the introduction of calculating machines and tabulators, the number had increased to 86. With the introduction of the tabulator, data processing started, and eight people were employed on this.

As the volume of business of the company increased, the number of employees also increased, and a more efficient means of accounting and data processing had to be introduced. By the end of 1964, 104 employees were in the Treasury Department and 18 employees in the data processing unit. About the same time the 1440 computer was installed, and the number of employees in the Treasury Department rose to 109 and in the data processing unit to 20. Current figures are 145 in the Treasury Department and 17 in the data processing unit. The figure for the data processing unit shows a reduction of three employees. Personnel reduction seems inevitable some time after the introduction of advanced technology, because with training and experience fewer people can handle the job. In this particular instance, the three persons were not actually laid off but transferred to other units. The reasons for picking these three individuals were not clear.

The above account shows that the introduction of advanced technology has not resulted in any employment attrition. Moreover, the airline growth rate in personnel of 11.25 per cent was less than the 12 per cent growth rate in the Treasury Department and the data processing unit. Nevertheless, the company feels that the growth rate in the Treasury Department and data processing unit would have been much greater if computers had not been installed.

It goes without saying that the primary concern of Ethiopian Airlines was to provide a quality of service that would meet international standards. From the point of view of the company, the use of advanced technology was not only desirable but essential, if the company intended to stay in business. The company considers too that it is rendering a useful service to the nation since it earns foreign exchange.

The company was aware of the negative consequences that advanced technology might have on manpower in a country like Ethiopia and decided to introduce new technology gradually. The two-stage introduction of advanced technology also helped in offsetting a possible negative reaction from employees.

Methods of determining wage levels

Ethiopian Airlines is one of the few Ethiopian organisations to have a scaled system of salaries and wages. Levels and responsibilities are categorised, and salary and wage levels are determined by the criteria that differentiate the job categories. While it would have been interesting to have a set of job descriptions in the Treasury Department and another set for employees who have been promoted to take up posts as computer programmers, supervisors and operators, this information was not available. Such material would have helped in determining whether or not greater competence and higher responsibility may have been the factors determining the award of higher salaries to the employees engaged in computer operation in the Treasury Department and the data processing unit. When the writer asked whether loyalty and good conduct were not also among the factors that were taken into consideration, there were no denials.

The company regards an assignment to computer operation as one that entails greater responsibility and requires greater competence. Accordingly, people who are given these assignments are raised a step or two higher. In general, those who are engaged in these kinds of jobs are better paid than the others. Even the girls who type in the Treasury Department and the data processing unit are paid more than typists in other departments and units. It might also be mentioned that these jobs entail higher status and greater prestige.

Another important influence on wage levels is competitiveness in the skilled labour market. It is imperative for Ethiopian Airlines to offer competitive salaries and wages in order to retain its trained and experienced workers. As an additional incentive, the airline has instituted a retirement benefit system. In fact, Ethiopian Airlines is the only non-governmental body that has such a plan.

GOVERNMENT EMPLOYMENT POLICY

Government employment policy gives priority to Ethiopians in all activities for which they are qualified. To this effect a public employment administration was created which brings together prospective employers and employees and thus makes available greater opportunities for Ethiopian nationals with the required skill.

Legislation exists to protect workers from arbitrary treatment by employers. This safeguard is embodied in the Labour Relations Decree, 1962. If an employee feels that he has been laid off without good cause, he has two courses open to him: he can appeal to his union to take appropriate action on his behalf, or he can appeal directly to the Labour Relations Board.

Although it is not made explicit anywhere, government industrial policy seems to favour labour-intensive undertakings rather than capital-intensive projects. A good example is the Winji Sugar Industry. There was considerable government resistance when the sugar estate proposed to introduce advanced technology in sugar processing. It also wanted to make extensive use of tractors and other machines in the cultivation and harvesting of sugarcane. The Government was against the proposal because thousands of people would be displaced, especially those seasonal workers who come to Winji from surrounding areas.

Annual leave, paid holidays, hours of work, overtime compensation and severance pay are also regulated by legislation. The main significance of this legislation lies in its contribution to industrial peace.

Since the Board of Directors of Ethiopian Airlines consists of ministers and other high-ranking government officials, government policy on employment matters is fully implemented.

There is nevertheless a difference of opinion between the airline managers and the Board of Directors about the speed at which advanced technology should be introduced. Management holds the opinion that the airline can very well use a higher degree of automation. But it does not press the issue vigorously. Apparently it means to achieve its goal gradually and by avoiding a direct confrontation with the Government. Considering the Government's attitude on employment, and considering the need for a business organisation to have an amiable relationship with a Government that can withdraw its support if its wishes are not accommodated, the current policy of the airline may well be the best one.

The fact that advanced technology has been introduced in stages has already been mentioned. The two-stage approach was probably adopted to enable the company to sound out feelings of both workers and the Government. As nothing happened on the completion of the first stage, the company felt that there would be little or no obstacle if it implemented the second stage.

CONCLUSION

At the beginning of this report, brief indications were given about the economic and social conditions that currently prevail in Ethiopia. It is worth while repeating here that unemployment is among the outstanding problems facing the country at the moment and this situation is unlikely to change for many years to come. Quite understandably, the Imperial Ethiopian Government is concerned with ways and means of improving the employment situation and is currently engaged in the nearly impossible task of reducing unemployment and underemployment in both rural and urban areas.

Only a few industries are likely to benefit from a greater use of advanced technology. Among the criteria to be observed when deciding for or against its introduction should be whether advanced technology will result in a large-scale growth of production and whether indirect benefits are likely in other areas. National industrial policy should also aim at encouraging industries with advantages in international trade. Income generated can boost investment and new industries, and help to achieve other developmental goals.

AUTOMATED PRODUCTION OF CANS IN TANZANIA

OMARI S. JUMA
Dar es Salaam

GUY ROUTH
University of Sussex, Brighton

The remarkable feature of this case is the smoothness of negotiation of problems which, in industrialised countries, have often been fraught with difficulties. Here, 21 years ago, the coconut palms were cleared, peasants recruited from the bush and a factory established employing high-speed machine production alongside a labour-intensive production line: the tinker's shears beside the machine press. Changes were introduced over the years until, by 1964, the main production lines were almost entirely automated.

Semi-literate workers, speaking only Swahili, have been trained as supervisors and can now do running repairs and maintenance, strip the machines and reassemble them. A machine shop has been established that can make all but the most complex parts.

It is true that, with the extension of automation and the advance of organisation efficiency, the company had, for three or four years, to carry a growing amount of surplus labour (a problem exacerbated by the great stability of the labour force). However, government, union and worker objections were in due course overcome and about 15 per cent of the labour force were dismissed in May 1968.

Opposition to redundancy followed familiar lines; but why has the process of innovation itself encountered none of the problems of industrial relations with which it is associated in industrialised countries? There does not seem to be any short answer. One can only assemble what seem to be relevant facts and offer plausible hypotheses. Comparatively high pay is one factor, the effort put into training is another—in particular, the *flexibility* built into the training programme, which incorporates the transferability of workers. Another important aspect may be the absence of vested interests in acquired skills which, in long-industrialised countries, automation renders obsolete.

In this paper some background material is presented relating to the Tanzanian situation, as well as information relevant to the company. From this, hypotheses will be drawn that may explain the comparative peace with which technical change has been introduced.

BACKGROUND TO INDUSTRIAL RELATIONS IN TANZANIA

Industrial relations in Tanzania have many features in common with those in other countries: life for most workers is hard, with a constant struggle to match income and expenditure; workers are constantly trying to lighten the burden of work, while managers strive to maintain or improve output; workers in any workshop tend to have common interests and to combine to promote them; leaders appear who are able to articulate hopes and grievances and present them to employers.

At the same time, there are some features that distinguish Tanzania from most other countries. The national income per head is in the range of the very poorest countries—less than US$ 100 per year. As one would expect, this is associated with a very large peasant population and a great army of unpaid family workers, as may be seen from table 16. (It should be mentioned that the figure of 717,000 employees includes guesses at the number of employees in peasant agriculture and in domestic service.)

The Central Statistical Bureau records employment of just under 350,000 in the modern sector in June 1967, while the results of the survey of industries for 1965 show that 45,000 employees were engaged in manufacturing establishments employing 10 or more people in that year. The average number of employees per factory was just under 80, and only 7 factories employed 500 or more.

One important contrast with the low-income countries of Asia is that there is an abundance of land in Tanzania: 341,000 sq. miles of land area with a population of 12,231,000 (1967), giving a density of just under 36 per sq. mile compared with about 300 in India. Only 6,000 sq. miles of this is desert, semi-desert or swamp. Thus the desperate poverty that shows itself in the towns of many poor countries is absent here. Nevertheless, there is an abundance of labour available for casual employment, and the average wages of adult male Tanzanians in manufacturing industry in 1967 were only 319 shillings [1] per month, in establishments employing 10 or more workers.

The average cash wage of Tanzanians for all industries at this time was 285 shillings and, low as this was, it was double the average of 1962. This represents a substantial rise in real wages, for in these five years the wage

[1] US$ 1.00 = 7.14 shillings (20 shillings = £EA 1).

Table 16. Distribution of manpower by status in Tanzania, 1967

Status	Numbers (thousands)	%
Employers and self-employed	2 420	35
Unpaid family workers	3 819	55
Employees	717	10
Total	6 956	100

earners' retail price index for Dar es Salaam registered a rise of only 13 per cent.

This increase in pay originated in government policy. Before the achievement of full independence, the Government had enacted, in 1961, the Regulation of Wages and Terms of Employment Ordinance, in terms of which a Territorial Wages Board was appointed in March 1962. As a result of their report a wage order was made in December 1962, prescribing the following minima: for Dar es Salaam, 150 shillings per month; for 18 main townships, 125 shillings per month; for all other areas, 100 shillings per month. This contributed to a rise of 33 per cent in the average earnings of African males between 1962 and 1963. These rates were raised by 20 shillings per month in July 1969, mainly to offset the results of increased indirect taxation.

Other legislation was enacted to improve the position of employees: the Severance Allowance Act of 1962 prescribed 15 days' pay for each year of service on termination of employment by an employer; the National Provident Fund Act of 1964 established a contributory fund to provide retirement bonuses; the Trade Disputes (Settlement) Act of 1962 provided machinery for the settlement of disputes and the postponement of strikes or lock-outs, and for the enforcement of the awards of an arbitration tribunal.

Of more direct application to the present study, however, is the Security of Employment Act of 1964. This Act, operative from 1 May 1965, provides for the election of a workers' committee in every concern employing 10 or more union members. The Act establishes a voluminous disciplinary code, optimistically aimed at covering all possible items of dispute between workers and employers and prescribing the appropriate remedy, and this the workers' committee administers in conjunction with the employer. But their duties include the promotion of efficiency and productivity (on which they must meet the employer at least every three months), and consultation with the employer on any impending redundancies and the application of any agreement on redundancies. If a dispute arises, appeal lies to a tripartite conciliation board and, beyond that, to the Minister of Labour. Thus the trade union (through its members) and the Minister are given considerable powers to control the activities of employers in the management of their staff.

We should also mention the Permanent Labour Tribunal Act of 1967, which replaced the Trades Disputes Act mentioned above. This requires the approval of the Permanent Labour Tribunal before any agreement between the trade union and an employer may be applied. This Act, designed to implement a national incomes policy, follows the recommendations of Professor H. A. Turner, reporting on behalf of the ILO in 1967. The Act prohibits strikes and lock-outs, pending a conciliation procedure, and unless the Minister has failed to refer the dispute to the tribunal in the specified period.

Act No. 18 of 1964 dissolved the Tanganyika Federation of Labour and its member unions and replaced them with a single union: the National Union of Tanganyika Workers. The General Secretary and Deputy General Secretary of the union are appointed by the President of Tanzania for a minimum period of five years, and the President also has power to dissolve the union and establish another in its place. This accords with government policy, which is to maintain close co-operation between the political party (the Tanganyika African National Union) and the trade union so that the latter, while promoting workers' interests, may also mobilise its members for the task of nation building.

In particular, the terms of reference of the trade union with respect to pay are given in Government Paper No. 4 of 1967, *Wages, Incomes, Rural Development, Investment and Price Policy*, which was based on a report by Professor Turner. Increases in wages and fringe benefits should not exceed 5 per cent in any year, provided that genuine payment-by-results schemes may afford earnings increases of up to 20 per cent for workers paid up to 150 shillings per month, a scale that is reduced to 5 per cent for workers earning 251 shillings or more. Increases should not result in price rises or redundancy, nor impede further industrial development.

Industrial relations in Tanzania are set in an intellectual environment very different from that of industrialised capitalist countries. In these countries social desiderata include the shortening of the working week, the lengthening of paid holidays and a general easing of the burden of labour; in Tanzania the tone is set by the Arusha Declaration [1] with its emphasis on self-reliance and hard work:

> Everybody wants development; but not everybody understands and accepts the basic requirement for development. The biggest requirement is hard work. (p. 14)

[1] This is a statement of policy of the ruling party, the Tanganyika African National Union, made at Arusha in February 1967. The page references are to *The Arusha Declaration* (Dar es Salaam, Tanganyika African National Union, 1967).

The second condition of development is the use of *intelligence*. Unintelligent hard work would not bring the same good results as the two combined. (p. 15)

Between *money* and *people* it is obvious that the people and their *hard work* are the foundation of development, and money is one of the fruits of that hard work. (p. 17)

The Declaration is written in simple language, intelligible to workers, and has great emotional appeal. Workers will, on occasion, work unpaid overtime to demonstrate their support. (Of course, this does not mean that they have somehow discarded their group interests in favour of the "common good". But there is not much intellectual support forthcoming for group interests that appear to conflict with the common good.)

THE METAL BOX COMPANY OF TANZANIA LTD.

The Dar es Salaam factory was established in 1948, as a branch of the Metal Box Company of East Africa Ltd. It was envisaged, at that time, that Tanganyika was to be an important source of edible oils derived, mainly, from groundnuts. The meat-packing industry was to be another substantial user of tin cans.

Expectations were disappointed as far as groundnuts were concerned. Production has expanded, but not on the grand scale then envisaged. But the production of oilseeds remains important and they are to an increasing extent being processed in the country. Meat and meat preparations come sixth in the country's list of exports, and great effort is being put into raising quantity and quality.

On 1 April 1966 the Dar es Salaam branch became a separate company with a local board of directors. It was a wholly owned subsidiary of the Metal Box Company of East Africa Ltd., which is in turn a wholly owned subsidiary of the Metal Box Company Overseas Ltd. in London.

This position was changed in 1968 when, after prolonged negotiations, a 50 per cent share in the Tanzanian company was acquired by the National Development Corporation. Although it is now regarded as a "parastatal organisation", this has occasioned no change in management or policy.

The company produces about 500,000 7-ounce and 12-ounce cans per week for corned beef products and about 150,000 cans for a new company processing edible oil. In addition, there are cans of various shapes and sizes for insecticides, fruit juices, paint, mineral oil and kerosene, miscellaneous powders, chemicals and patent medicines, as well as advertising display plaques and lapel badges.

Departments

Administrative Department. This department includes the accounts section and all other administrative branches, such as personnel and surgery.

Engineering Department. This department is responsible for the maintenance and efficient running of all machinery in the factory, and is fully equipped with the necessary tools. It is rarely necessary to have repairs done outside the factory. The department is responsible for all repairs to machinery, vehicles and buildings. Training has been a special problem here, for the skills required range from welding to electronics (see "Training" below).

Printing Department. Here, plain sheets of tinplate are decorated at high speed. The work is highly specialised and requires a high degree of skill. The labour force in this department has been stable for a number of years now, partly because jobs in it are regarded as valuable prizes and partly because of lack of openings for these skills outside the company. As a result, the workers form an efficient and closely knit team. Three categories of employee are distinguished: highly skilled, semi-skilled, and specialised labour (see "Manpower" below).

Production Department. This consists of a stores department, feeding and being fed by three production lines. The "General Line" produces a wide diversity of products and is divided into six subsections according to product. Most of the subsections are semi-automatic, with products fed into and out of machines by hand and being subjected to various manual processes. Cans for insecticides and oil (ranging in size from 1 pint to 1 gallon, but including some 2.5-gallon and 28-pound round drums) are produced at speeds varying from 250 to 500 per hour.

Paint cans are produced in sizes varying from 0.5 pint to 1 gallon, again at speeds between 250 and 500 per hour. Four-gallon *debes* are made for the oil and cashew-nut industries at a rate of 800 tins per hour. Semi-automatic power presses turn out components for other parts of the General Line, and bench, hand and foot presses are used for small components and items such as badges and bicycle licence discs. Workers employed are listed in table 17.

While these 120 workers turn out about 1,000 cans an hour on the General Line, 31 workers on the two "Open-Top Lines" [1] produce round or rectangular cans at a rate of about 26,000 per hour. The second Open-Top Line was introduced in 1964 with a capacity of 14,000 per hour when the

[1] An open-top can is one that must be opened by a can opener or key.

Table 17. General Line workers, Metal Box Company of Tanzania Ltd.

Job	Number
Key welders	7
Irregular die line	3
Lap seam	7
General Line	3
Slitters	7
Round built-up line	15
Irregular built-up line	20
Presses	25
Four-gallon line	33
Total	120

other was adapted to produce round cans at a rate of 12,000 per hour. Hourly output per worker is 8.3 cans on the General Line and 871 cans on the Open-Top Lines.

A sample automated process

A variety of the open-top processes have to be carried out at high speed and with great accuracy. Automated machines that are not set accurately will cause considerable trouble during subsequent operations. By way of illustration, we give a breakdown of the operation of the "4B Bodymaker" and "54 BS Side Seamer", which form the body and solder the side seam of rectangular cans used for packing corned beef.

Note that only one worker operates the Bodymaker and that one other inspects at the far end, checking cans for weak laps, solder penetration and other defects. On the semi-automatic General Line, several times as many workers are needed to make machines execute these operations.

The 14 operations are as follows:

1. Blanks are fed into the hopper and are drawn in by rubber suckers and blank separators. The auxiliary fingers then take the blank to the body stops and the blank is clip-notched on the right-hand corners and slit-notched on the left-hand corners.

2. The blank is pushed to the body stops, and hooks are formed at right angles.

3. The blank is again pushed to the body stops, and hooks are formed to 28° with the help of folder steels moving in an arch.

4. While being transferred by the rear pair of fingers on the main feed bars from this station to the next station, the hooks and laps are coated with liquid flux in readiness for soldering.
5. The blank is received at this station and registered by body stop and side supports. It is then clamped between the expanding mandrel and anvil steel. Two forming wings come down and form the blank oblong on the oblong expanding mandrel.
6. The mandrel is expanded, making a tight interlock of the hooks, and the bumper steel comes up, locking the hooks and making a tight side seam. The oblong can is then transferred to the next station by the extractor bar.
7. The oblong can is picked up by the conveyor chain and carried along to the front can guide.
8. The oblong can enters the solder horse and passes over the solder roll. The solder roll, revolving in the solder bath, transfers solder from the bath to the side seam of the oblong can.
9. The can passes over the pre-wiper burners, the solder splash-reduce and the wiper unit. Heat from the pre-wiper burners maintains the solder in a molten condition so that excessive solder on the side seam can be removed by the wiper mop with the minimum of pressure.
10. After the side seam has been wiped and before the can body leaves the solder horse, cool air is blown on the side seam to solidify the solder as quickly as possible.
11. The can body, still on the conveyor chain, leaves the solder horse and enters the rear can guide. At this stage further cooling is effected.
12. The can body leaves thc rear can guide and is extracted from the conveyor chain by the extractor chain which delivers it on the cooler chains.
13. At this station, still further cooling is effected while the can is carried along to the can turnover unit.
14. The can is turned through 90° by the can turnover unit, then rolls down into the vertical elevator. The vertical elevator lifts the cans and pushes them into the runway, thence feeding them to the flanging machine.

MANPOWER

Job evaluation

The factory operates an elaborate job evaluation scheme, in terms of which each job is broken down into 12 factors and graded according to the

Table 18. Job evaluation, Metal Box Company of Tanzania Ltd.

Group	Category of worker	Range of points	Workers included in category	Number employed per group in 1969
6	Unskilled	Up to 60	General labourers, feeder operators, office messengers	46
5	Semi-skilled	61 to 75	Labourers with specialised tasks, hand solderers, machine operators (39 jobs are distinguished)	129
4	Skilled	76 to 115	Watchmen, quality measurement line inspectors, artisans' assistants, machine operators requiring higher skill than those in group 5 (10 jobs are distinguished)	36
3	Artisan, grade III	116 to 170	Fork-lift truck operators, lorry and van drivers, machine setters	25
2	Artisan, grade II	171 to 205	Artisans more experienced than those in group 3, able to work with little supervision	8
1	Artisan, grade I	206 and above	Coater operators, printers (metal decorators), artisans able to locate and rectify faults and direct activities of grades III and II artisans	30

total of points scored. According to its score, it is then allocated to one of six groups, as shown in table 18.

Departmentally, the labour force was distributed as shown in table 19.

One of the outstanding characteristics of the labour force at Metal Box is the negligible rate of labour turnover. A job at Metal Box (or at any of the big, modern plants in Tanzania) is an asset not lightly to be surrendered, so that labour turnover is extraordinarily low as compared with that in Britain or the United States. This stability of employment has many advantages to employers and workers, but does impose certain difficulties in the way of technical advance.

By the beginning of 1968 the position as regards the numbers employed was not much different from that in the first part of 1964 (356 in May 1964; 343 in January 1968). Output was given a certain flexibility, however, by a usage common in East Africa: the employment of casual workers. Thus employment had risen to 435 in April 1965, but it was possible to reduce it to 383 in June of that year by the dismissal of casual workers. This operation passed off peacefully because it had been known that the additional employment was temporary pending the transfer of the business of an oil company from Dar es Salaam to Nairobi and the introduction of an automatic flanger.

There was a further reduction from 344 in April 1968 to 259 in May 1968, followed by stabilisation at just under 300 in June. This is especially relevant to the present study and will be described below.

Table 19. Staffing summary, Metal Box Company of Tanzania Ltd., 1969

Staff	Number
Executive	4
Production planning	4
General services	15
General Office	19
Despatch and warehousing	13
Security	12
Stores	20
Production clerks	4
Quality control	10
Engineering department	24
Printing department	14
Open-Top Line	31
General Line	120
Relief labour	2
Total	292

The employer finds his initiative restrained by the power of the trade union, buttressed by the workers' committee and the Security of Employment Act (see above). It is a tenet of the official incomes policy, too, that increases in pay should be offset by increases in productivity but should not result in redundancy, and this creates difficulties for plants for whose products demand is not rising.

Pay

Pay in the big firms has traditionally been well above general rates. Metal Box was put even further ahead in April 1966 by an arbitration award giving an increase of 25 per cent over the minimum wage. This had the effect of increasing the total wage bill of the company by 17.5 per cent. All increases were backdated to 1 January 1966. The size of the award came as a shock to the company, and the chairman took the opportunity of writing to the Minister of Labour on 3 June 1966 to remark: ". . . while we have always done our utmost in the past, and shall continue to do our utmost in the future, to protect our customers' interests by even greater efficiency of production (which might ultimately lead to some reduction in the number of persons employed), it is inevitable that an increase of this magnitude must lead to some upward review of the company's prices". However, over the succeeding months much work was done to find ways and means of

Table 20. Minimum wage rates, Metal Box Company of Tanzania Ltd.

Group	Category of worker	Shillings per month
6	Unskilled	255
5	Semi-skilled	283
4	Skilled	335
3	Artisan, grade III	360
2	Artisan, grade II	450
1	Artisan, grade I	750

reducing costs, and as a result the company was able to maintain its prices at the original level.

A further increase was effected by a voluntary agreement signed on 16 November 1968. This, combined with some regrading, increased the average wage from just under 327 shillings to 339 shillings per month—an increase of 3.8 per cent. The minimum rates specified under the agreement are shown in table 20.

In 1967 the Central Statistical Bureau collected information relating to 232,000 adult male citizen employees. Their frequency distribution by range of pay is shown in table 21.

The exclusion of agricultural estates from the calculations would cause a substantial upward shift, for their 124,000 workers averaged only 161 shillings in 1967, while the 31,000 in manufacturing averaged 319 shillings. But the point is that workers in manufacturing are a small, favoured group, of which the employees of Metal Box are themselves in a somewhat favoured position. Hence the great stability of the labour force and the difficulty of reducing it with technical advance or business retreat.

Training

Since the company's formation in 1948, its employment policy has been little changed. Tanzanian Africans are given preference, and only when the necessary technical qualifications, experience, integrity and industry are lacking is the position considered open to a local person of a different race or an expatriate. Since 1961 expatriate experts have been employed only where a suitably qualified national cannot be found. The original number of 10 expatriates has been reduced to three. Training is inevitably a slow and continuous process, but it has always been the aim that the company shall eventually be staffed almost entirely by qualified and experienced Tanzanian citizens.

Table 21. Frequency distribution of employees by range of pay, Metal Box Company of Tanzania Ltd.

Shillings per month	%
Up to 99	6.0
100 to 149	22.4
150 to 199	29.3
200 to 299	18.7
300 to 399	9.3
400 to 499	3.9
500 to 749	5.5
750 to 999	1.9
1 000 and above	3.0
Total	100.0

When the company began operations in East Africa, the emphasis had to be placed on on-the-job training. There is no fund of craftsmen available with wide experience in maintenance or machine-tool setting or operation, whose skills can readily be adapted to a new line. Few craftsmen have even served an apprenticeship, so that their range of skills and adaptability is small. The *fundis* who do the work of craftsmen will have learnt their job by watching others; their theoretical knowledge will be small; they may be illiterate and not easily trained for unfamiliar skills.

Thus, when the company began operations, it was necessary to engage workers who were either unskilled or, at best, semi-skilled. Initially the company endeavoured to train craft apprentices, following the pattern which at one time was universal in the United Kingdom. Candidates were selected according to their technical aptitude and allocated to trained journeymen for instruction. Unfortunately this system had serious deficiencies, both in the United Kingdom and in East Africa. As a result, the Metal Box Company of East Africa Ltd. decided to set up a fully equipped apprentice training centre, supervised by a qualified training officer. An expatriate, with wide experience both as a mechanical engineer and as a training officer, was appointed to this post in 1963.

His initial task was to make a survey of training needs. This was done through "Training within Industry", pilot courses and close observation in the various factories. The results revealed a persistence of low standards (especially in educational levels) in the workforce and a continued shortage of skilled personnel.

As a result of these findings, it was decided to establish the Apprentice Training Centre at the Thika factory in Kenya, which would cater for students from all parts of East Africa. This centre, well set out and fully equipped, came into operation on 1 January 1966. The cost of running the centre was £8,000 in the first year, of which the Tanzanian company paid about £3,000. It is currently running at £20,000, of which the Tanzanian company pays half.

In general, the craft and technical schemes have worked well. At the end of 1968, Mr. W. T. Jones, the head of the Apprentice Training Centre, reported on two trainees:

> Messrs. N. Tibursi and Reuben Lazaro joined in January 1967 from Moshi Technical School, having already passed Part I of their City and Guilds craft practice course. They have recently sat their final examinations. Due to the progress they have made, I recommend that their apprenticeship period be reduced from five to four years. . . . They are both good at practical work and their Polytechnic teacher is very confident that they will obtain good passes in their final City and Guilds exams.

These two, who duly achieved the expected examination successes, are now back in Dar es Salaam, doing two years' training in the toolroom on the production lines.

On a third trainee Mr. Jones reported:

> A good, all-round craftsman. Good at general machining including turning, shaping, milling and grinding. Good at fitting with a high degree of accuracy. He is very interested in any job he undertakes and is always eager to please, and is capable of working on his own. His Polytechnic reports are very good. Potential toolmaker. At the moment, he is in the running for the "Apprentice of the Year" award.

We include these reports to indicate that the problem of machine maintenance is being tackled successfully despite the absence of an engineering tradition and a stock of craftsmen in Tanzania.

It is noteworthy that the two factory foremen at Metal Box, although men of limited education, have mastered the operations of setting and routine maintenance of the machines involved. They can tell at a glance if all is not well and can readily diagnose faults and rectify them (unless, of course, it is a job for the toolroom). They can dismantle and reassemble the machines, despite their complexity.

A considerable advantage enjoyed by the company is its freedom from the tumultuous labour history that has shaped industrial relations in economically advanced countries. In earlier stages of industrialisation, the machine is used as an adjunct to manual skills. Vested interests in these skills are established and this creates an opposition to technical advance that manifests itself in manning disputes and demarcation disputes.

Here, by contrast, there were no vested interests, and the company was able to require and obtain complete flexibility with regard to its production lines. Its policy is to train all process workers so that they can perform any jobs on the production lines and, in practice, to transfer them from job to job on a predetermined rota. This applies to job nos. 4 to 42, which make up group 5 in the job evaluation scheme (table 18).

This process is facilitated by the absence of competing trade unions for, as we have seen, in Tanzania there is only one trade union. Of course, this would not automatically exclude demarcation disputes if workers of one occupation objected to the introduction of workers of another occupation, but it would prevent institutional support being given to such a dispute.

Industrial relations

This acceptance of change and flexibility is not at all due to quiescence on the part of the workers. Although the company has enjoyed industrial peace, this has occasionally been disturbed. On these occasions the workers have manifested the same sort of solidarity, determination and ingenuity that is associated with direct action by workers in industrialised countries.

There was, for instance, a "go-slow" that lasted from 28 March to 5 April 1966 in protest against certain features of the company's medical service. This was ended only after the Labour Commissioner had undertaken to arrange a commission of inquiry into the complaints.

Similar disturbances followed the appointment of a foreman from outside the company, when meetings of the workers' committee and lunch-hour meetings of workers were followed by a "go-slow". In spite of vigorous opposition, the company stood firm on the appointment, and the foreman turned out to be very effective.

These cases are cited to demonstrate that the lack of resistance to innovation is not due to the docility of the labour force.

AUTOMATION, PRODUCTIVITY AND REDUNDANCY

Output per worker

This changes considerably from month to month, the greatest variation being in the demand for cans. This, again, varies with the supply of products for canning—principally with the supply of beef to Tanganyika Packers. With highly automated lines, it is an easy matter to put on an extra shift, or, on occasion, an extra two shifts. An additional shift involves only 23 workers.

Continuous efforts have been made to obtain the most economic use of raw materials, since the tinplate used for can manufacture constitutes the major item of expense—greater, for example, than the labour costs. At the same time, efforts to economise in the use of labour have not been neglected. The correct allocation of labour to a given line is of prime importance. Many years of experience in operating the factory have proved that, if more men than necessary are working on one line, not only are there additional costs for wages and other benefits, but in most cases the additional man will get in the way of his colleagues and thus reduce production efficiency as well.

The company knows very accurately the amount of labour required in any section for optimum efficiency. The aim is to have sufficient labour to give every man a full and fair day's work. Since none of the lines is free from interruption caused by irregularity of demand, the policy is for mobility of the labour force, so that each man is proficient on two or more lines. Although there are many unskilled or very simple semi-skilled jobs on all lines, by and large the policy is to keep the labour force within three main production areas: printing, Open-Top Line and General Line.

Much work has been done as regards factory lay-out and other general improvements: for example, putting machines closer together to save carrying components from one operation to another, or making room for fork-lift trucks to carry one big load rather than having individuals carrying many small ones. We may also mention a considerable increase in the use of pallets for storage, which has saved much handling of individual cans and led to far greater efficiency.

The value of output per head averaged £186 in 1966/67, £179 in 1967/68, and £206 in 1968/69. The low was £94 in December 1966 and the high was £348 in August 1968. Prices did not increase during these years. The variation is mainly due to the company's inability to reduce employment in proportion to demand in slack months.

Innovation and redundancy

The company has been singularly free of resistance to innovation. It may be inferred *a priori* that this would be a feature of most countries that begin industrialisation in the era of automation, for, as mentioned above, they lack the process (production) craftsmen who are such an important feature of industrially advanced countries. Thus there is no large, highly organised body of workers whose skills are made obsolete by the extension of automation.

It is worthy of note that the workers at Metal Box have accepted technical innovation without demur. It does not seem to cross their minds that it is in any sense a threat to their jobs. It is true that when the new Line No. 1 was installed the labour force was not reduced—in fact, two extra workers were engaged. But of course this is not enough to mollify workers in industrially advanced countries, whose morale seems prone to fall as soon as plans for innovation are discussed.

But while Government, trade union and employers are all agreed on the urgent need for raising productivity in Tanzania, there remains an unresolved contradiction between official and trade union policy. Professor Turner and the Government (in Government Papers Nos. 3 and 4 of 1967) call for a vigorous effort from all concerned to raise the productivity of labour—but at the same time they want this to be done without loss of jobs. This means, in effect, that productivity should be raised no faster than the demand for the product. A 10 per cent rise in productivity with a less than 10 per cent rise in sales would of necessity produce a surplus of labour. But the demand for cans is a derived demand, depending on the demand for the products of its customers. This will undoubtedly expand with the growth of the economy and exports, but it has failed to expand fast enough to eliminate an accumulation of surplus labour at Metal Box. At the same time, Metal Box has been expected to keep down the price of its products, despite the rise in the price of tinplate, as an aid to the sales (especially export sales) of its customers.

By 1966 the problem of surplus labour had already appeared, and the company began seeking ways of reducing its workforce. However, the obstacles appeared to be insuperable, and the surplus labour simply had to be carried. In April 1968, however, nature intervened. Torrential rains washed away bridges, made rivers impassible and turned roads into seas of mud. Events of this sort are not unexpected, but on this occasion railway lines were inundated or swept away as well, so that the national transport system came to a standstill. Not only did the supply of beef to Tanganyika Packers stop, but it was not possible to send out cans to up-country customers.

The management met the workers' representatives on 2 May 1968 and reported to them that they were faced with the immediate necessity of laying off 97 men. They offered the workers a choice of three alternatives:

— termination of employment, with one month's pay in lieu of notice, severance pay in terms of the Security of Employment Act, leave entitlement, plus their entitlement in terms of the company's provident fund;
— lay-off without pay, but with guaranteed re-employment when things returned to normal, possibly after two or three months;
— some combination of the above.

The union officials addressed the 97 workers concerned at a meeting in the canteen, and the unanimous choice was for the first option. The company accepted the suggestion that these workers be given first choice of re-employment when the crisis had passed.

This cost the company about £5,000 and gave the redundant workers about 1,000 shillings each—equivalent to nearly four months' pay. Those unable to find other factory jobs would thus have enough to begin farming a few acres of land, or even to set up a village shop (*duka*).

The 97 were selected on the basis of "last in, first out" for each section. Their year of engagement gives a good indication of the remarkable stability of the labour force: 13 of those dismissed were engaged in 1960, 13 in 1961, 42 in 1962, 22 in 1964, and 1 in 1965.

Consolidation

Once the transport crisis was over, the company planned to bring its workforce up to about 300, the optimum figure for efficient production. The case was argued with the union on 12 and 13 June 1968, in conjunction with the union's demand for an increase in wages.

National incomes policy allowed a maximum increase of 5 per cent per year, provided that the productivity of the workers concerned rose by not less than that percentage. In fact, with a workforce of 300, the company estimated that productivity in 1968/69 would show a rise of 11 per cent over that in the previous year. Thus a wage increase of 5 per cent would create no strain on labour costs. Concurrently, the company proposed to introduce a job evaluation scheme.

At first the union argued that it was not in the interests of the country's economy that labour should be reduced in order to increase wages "through artificially increased productivity", but finally agreed, in the following terms:

— that the parties should recognise the desirability of maintaining the labour force at a level compatible with the increase in productivity as outlined in government policy . . . on wages, incomes and prices; and that both parties should recognise and accept that in accordance with the company's 1968/69 forecast, a labour force of 300 workers should be maintained to give rise to an 11 per cent increase in productivity over and above the 1967/68 figures;

— that the union, in collaboration with the management, should undertake to educate the workers to recognise and accept the fact that without rises in productivity it would not be logical to demand increases in wages: they intend to do this by improving the workers' attitudes through audio-visual aids and also through arranging discussion groups on how to achieve and increase productivity.

This agreement was in due course approved by the Permanent Labour Tribunal and was signed by the parties on 16 November 1968. It laid down the groupings and wage rates already described above, and stipulated that all current employees on the minimum wage of 255 shillings would receive an increase of 15 shillings per month, whilst those getting more than this would be given an increase of 10 shillings. It was further agreed that no redundancy was likely to occur and that the company's selling prices would not be affected.

CONCLUSION

At the beginning of this case study we observed that, while opposition to redundancy followed familiar lines, opposition to technical innovation was singularly lacking. This seems to be characteristic of East Africa generally and may perhaps be a general feature of countries in the early stages of industrialisation. In this connection, we listed the following factors:

1. The pay of those employed by industrial firms is comparatively high. This leads them to place a high value on their jobs and to be unwilling to take action that may jeopardise them. Obversely, a change introduced by management that is not seen as a threat to job security will not be resisted.

2. The company has done most of its own training and is known to be willing to retrain workers when the need arises. The investment in its employees' skills gives the company a vested interest in retaining them.

3. Of greater importance is the flexibility on which the company has insisted from its inception. Workers are periodically shifted from process to process so that they do not establish vested interests in particular jobs. Thus they do not feel that it is "their" job that is threatened by the installation of a new machine.

4. In long-industrialised countries there are large numbers of workers who have acquired manual skills as machine operators or non-manual skills as clerks or book-keepers. These are the classes who feel themselves threatened by the move towards automation and from whom resistance to automation originates. High rates of labour turnover, again, reduce the workers' sense of identification with any particular firm. In Tanzania these classes of workers with vested interests in the preservation of old techniques are almost completely absent. Established employees are given considerable security of employment, whilst casual workers have so little that they expect none.

5. One should mention, too, the sense of limitless wonders flowing from the industrially advanced countries. This is accepted as both desirable and inevitable, so that it would be pointless to resist it.

LESSONS FROM THE CASE STUDIES

W. PAUL STRASSMANN
Michigan State University

Even when a statistical universe is fairly homogeneous, one hesitates to base generalisations on a sample of six. A single exception to a statement technically puts it below the 90 per cent confidence level. Our set of six "automating" firms is not even a random set, and it comes from a very heterogeneous universe of unknown size. To define the size of the universe would have meant taking on the dubious exercise of determining just how much advanced technology makes a firm "automated".

Nevertheless, if generalisations cannot be cut out altogether, six cases with a common approach are vastly better than a casual flotsam of information. If the six cases can speak for the less developed world as a whole, they are indeed worth adding up.

The nations from which our six cases come are heterogeneous enough to represent the Third World. Some have alarming inflation, others have monetary stability. Some are among the world's poorest, others have regions (like southern Brazil) that easily approach Mediterranean European levels of living and productivity. Population sizes range from millions to hundreds of millions. One was a colony a decade ago; another has been independent for 15 decades; and a third, Ethiopia, lost its independence only temporarily in the 1930s. Not all have been politically stable in the past two decades; and, to put it mildly, the forms and nature of their government differ widely.

If there are variations among the six cases of automation, do these national differences account for them? A review of the cases suggests that such differences did affect the response of government and labour in a few instances but that most variations came from size and sector. World-wide generalisations may therefore not be safe, but at least they are not obviously irrational. Here we shall first compare firms in general; next, their type of automation—the *what*, *why* and *how* of introduction; and finally, the reactions of labour and government.

CHARACTERISTICS OF THE SIX FIRMS

For convenience of exposition, let us refer to the six cases by initials, noting that the Dacca and Chittagong case covers not only the United Bank and the Habib Bank but also 20 of their computer clients (see table 22).

Age and size

Diversity in age and size among the six cases is great. BB and X go back to the nineteenth century, MB and EAL date from the 1940s, but PR was started in 1953 by a husband, wife and maid in a kitchen. In the late 1960s PR employed 230 and MB 300, while BB employed 42,500. None was small, but BB was a giant. The annual volume of PR came to US$ 1.4 million in 1969, MB produced US$ 2.1 million, while the operating revenues of EAL were over US$ 20 million.

But this diversity is deceptive. Compared with others in its economic branch, each firm was old, usually the oldest. PR and X entered a field of handicraft competitors; MB and EAL were sole producers. Five of the firms were the largest (or only) producers in their field, and the remaining anonymous X was "one of the largest".

Three of these largest and pioneering firms—MB, EAL and BB—began introducing automation or computers in 1964. The others followed: X in 1965, P in 1967, and PR in 1969. We are therefore witnessing a process in its earliest stages. Lagging 10 or 15 years behind the industrialised countries, the largest and best established firms of developing countries are beginning to adopt computers. Anyone familiar with technological diffusion will expect that smaller firms in each branch will follow, that following will be easier than pioneering, and that the spread of advanced technology will increase with the coming of smaller, cheaper and more versatile computers and electronic devices.

Nature of the advanced technology

Advanced technology can overhaul a firm's entire operation or merely change a part of its product lines or activities. These can be existing or new activities. The new activities can be simple growth along familiar lines or expansion into lines novel for the firm.

The overhaul of part of an existing activity was by far the most common effect of computers and automation in our six cases. X, EAL, BB and P all had to keep track of costs, payrolls, inventories and accounts before they ever had a computer. MB already had a production line for producing

Table 22. List of firms examined

Firm	Product	Location	Abbreviation
Banco do Brasil SA	Commercial banking	Brazil (various cities)	BB
Productos Ramo SA	Cakes and biscuits	Mosquera, Colombia	PR
Company X	Cotton textiles	India (various cities)	X
United Bank, Habib Bank and clients	Computer services	Dacca and Chittagong	P
Ethiopian Airlines	Air transport	Addis Ababa, Ethiopia	EAL
Metal Box Company of Tanzania Ltd.	Cans	Dar es Salaam, Tanzania	MB

open-top cans in 1964, when a second line was added and both were overhauled. PR's machines for weighing dough or breaking eggs at 100 per minute certainly took the place of old activities, but the biscuit line was new. Also new was X's use of the computer for blending cottons and for identifying the most profitable product lines. If one considers the branch banks of BB as separate units, one may consider 88 of them to be cases where an entire operation was overhauled, most conspicuously in the roles of the bank tellers.

Every one of these adoptions involved foreign private technical assistance. MB and EAL were closely associated with foreign enterprises, and employed expatriates. X, BB and P had the assistance of the computer companies, and PR had help from sellers of foreign machinery. PR had the least assistance, but it also undertook the least remarkable technological advance.

Economic effects for the firms

For these innovating firms, the technical features and origins of advanced technology were just the means toward one economic effect: higher profits. But higher profits could come about in a number of ways. For example, there could be an absolute fall in costs. On the other hand, costs could rise but volume and revenue could rise even more. Finally, sales and income could increase through better quality.

If there was an absolute or relative decline in costs, two further questions arise. First, was a reduction in labour costs the main target of the new technology? Second, was the elasticity of capital/labour substitution high enough to affect the choice of technique through small changes in relative factor prices?

There can be no doubt about it: a slowing-down in employment growth, and hence a slower increase (if any) in the payroll, was the most common economic effect (P, BB, MB and EAL). In X, improvement in the choice of cotton blends and products lines outweighed savings in labour. The expansion of the X computer centre's staff even matched the lower need for clerks elsewhere. Only in PR was any labour-saving effect of a new technology relatively incidental to better quality and to the volume-increasing effect on a new product with a new production line. Fresher cakes and a novel line of biscuits were the purpose of new equipment. But even PR introduced packaging machinery that displaced workers who had to be reabsorbed elsewhere in the plant. With all these labour-saving changes, only MB actually reduced its payroll.

IMPLICATIONS

Capital/labour substitutability

A close look at the cases suggests that the elasticity of capital/labour substitution was high and that management adopted new technology when it was barely better than marginal. This observation does not impugn the quality of management—on the contrary, it testifies to alertness and rationality. What may have been irrational from a social point of view were factor prices that did not keep capital-intensity submarginal.

Let us survey the evidence on the high elasticity of substitution and the bare super-marginality needed by management to substitute. In the PR packaging machine, raised capital costs almost matched lower labour costs, and only the better appearance of the thermosealed polyethylene package made the change look profitable enough.

EAL adopted the 029 five years before it expected volume and labour savings to justify the expense. In the meantime, only the uncertain advantages of greater speed and automatic data interpretation supported the investment.

During 1967-69 BB modernised 259 manual branches and established 39 new ones. But electronic data processing was provided in 47 branches only, while 251 were introduced to mechanical calculators. The more labour-intensive calculators were obviously still competitive with electronic data processing in the majority of branches. BB seems to have been particularly creative in introducing systems of rationalisation and simplification that raised the productivity of non-electronic methods.

The experience of BB contrasts sharply with that of P. Of the 20 computer-service clients of the two banks in P, 17 went straight from manual methods to electronic data processing, skipping mechanisation altogether. Since they were using excess capacity of the computers, a waste of capital is not necessarily implied, although salaries in Dacca were only about one-third of the Brazilian level. The marginality of their decision is suggested by the inducements that were apparently needed to gain trial. At first, the computer-owning banks charged no more than Rs 250 per month to "pay for paper". Later the charge went up to Rs 400-500 per hour. This amount was about one-third less than the wages of released clerks. Problems with computer breakdowns, spare parts and a possible need for stand-by generators may cause a rise in the hourly rental. In such organisations as the Dacca Water and Sewerage Authority, manual methods lacked efficiency in pressing collections primarily because of poor advance planning for middle management as operations grew. Better planning and training can therefore raise manual productivity. In the light of all these cases, it therefore appears that the elasticity of substitution between the computer and manual methods will remain high for a number of years.

In case X, computer processing of statements on production, loom efficiency and payrolls cost about six times as much as manual processing. Only better identification of the most profitable inputs and outputs could justify the installation. It is not clear that this identification had to be a continuous process, demanding ownership of a computer. The Indian XYZ Bank, the largest commercial bank, described by Dayal in his comparative study on "Preconditions for Effective Use of Automated Technology in India", found that the computer was necessary for balancing inter-branch accounts and statistical reports but that, otherwise, continuing with manual methods was entirely feasible.

Manpower effects

If highly advanced technology was only marginally superior to alternative methods for management, whilst at the same time seriously reducing employment growth for society, the policy implications seem obvious. However, we should first survey in more detail the employment and manpower effects in the six case studies.

It is curious that labour-saving equipment can be most easily introduced in economies where employment is most desperately needed. In these economies wages are likely to be very low and the wage bill a small percentage of total costs. Hence management will be less hostile to carrying a few extra workers in the short run, until attrition or output expansion re-establishes a desired ratio. Their willingness to wait lubricates and disguises basic labour displacement. Winning job security will seem a far more precious and adequate achievement to workers in a country with high open and disguised unemployment than to those in a fully industrialised welfare state. Personal benefits take a heavy priority over indirect implications.

These statements may be supported with random data from the six cases. The share of the wage bill in output was only 4.5 per cent in PR and 7.5 per cent in MB. Typical monthly pay (converted to US dollars) was $51 for the mainly unskilled workers of PR in Colombia; $48 for the average worker of MB in Tanzania; and $36 for the average lower division clerk in Dacca. All the firms claimed to be paying wages substantially above the average for their sector. Workers transferred or otherwise affected by modernisation were given pay increases ranging from one or two steps upward in the usual salary scale (X, EAL) to increases of over one-fifth (MB) or one-third (P).

In four cases (MB, X, BB, P) the new technology created a surplus of workers that was simply absorbed at the employers' expense. Where mechanical calculators were already established, computers meant a surplus of 30-40 per cent of the clerks; where they were not, the surplus was more than half (X, P). In all these cases, employment growth virtually ceased. Between 1966 and 1969 BB set up 39 new branches but employed only 807 more people. In 1966 employment (in the field or at headquarters) averaged 65 persons for each branch operation. If that average had been maintained, some 1,700 additional persons would have been employed in 1969. Had 1,700 jobs been lost? Perhaps the number was smaller because new branches are smaller. Perhaps it was bigger because some of the newly hired 807 workers did not service the new branches.

Only PR and EAL did not find themselves with an excess of workers after modernisation. PR was growing fast enough to need surplus workers from one production line immediately on another. EAL found that it could not operate its Treasury Department with fewer people after bringing in the computer, but future staff expansion was definitely forestalled. MB was able to reduce its employment by 22 per cent in connection with a disruption by floods, but afterwards the number of permanent positions was rigidly fixed.

Training

The advanced technology required more trained operators, and in this sense of quality replacing quantity the reduction in labour inputs was less than might appear at first glance. At the same time, inequality rose. Jobs were better; job opportunities, worse. Being in was better; being out was more likely.

Except for P, the firms mounted their own training programmes, usually with foreign assistance. In P the firms that sold the computer services trained the necessary personnel of their clients. For its own staff the United Bank also varied from the pattern by mainly bringing fresh graduates, rather than existing employees, into the training programme. The other cases mainly showed selection of present employees for retraining. Not surprisingly, it is in P that the trade unions appeared weakest and most discouraged by both government and labour.

Trade unions and the response of labour

Although union activity may have been weakest in P, nowhere was it strong. PR had an unaffiliated internal union that could not even persuade its members to insist on insiders for vacancies in supervisory positions. X could introduce automation without consulting its clerks' union in advance. MB's union was consulted before steps were taken towards modernisation, but it was given no voice in promotions or in the organisation of training programmes. This union had been capable of organising "go-slows" over relatively minor issues (for example, certain features of the medical service). In 1964, however, the union was integrated into the political structure and necessarily became less independent. Neither unions nor workers in any of the cases objected to any technological change that avoided dismissals. The negligible amount of turnover is also consistent with a sense of relative satisfaction on the part of the workers. Only in X did the transferred workers complain—first, because they had too much to do, and next, because they had too little to do and feared further displacement. In none of these cases had workers developed a sense of vested right to a particular post. In MB, where the union was probably strongest, training and reassignment policy was deliberately designed to prevent that sense of post ownership.

Government

More diversity exists in the role of government in the six cases than in any other general factor. Except for some subsidiaries of X, management was uniformly eager to adopt novelties on the slightest pretext. Labour was uniformly ready to accept anything that cost no current insider his job. But the government response varied from ignoring, to opposing, to demanding technological change.

Official Colombian documents and speeches profess concern about the costs and gains for employment from advanced technology, but the case of PR was apparently too insignificant to rate government attention. Government had a chance to play a part at the time of granting foreign exchange, but apparently no guidelines or machinery for employment promotion existed at that point. In effect, PR was ignored.

By contrast, EAL felt that government was hostile to modernisation, as suggested by its refusal to allow mechanisation of some sugar plantations.

EAL's management wanted to extend the use of its computer, but feared adverse reactions from labour tribunals and the government ministers on its own Board of Directors. Nevertheless, the management seemed to feel confident that a strategy of gradual expansion of electronic data processing could work.

In MB, government practically insisted on technological advance. It wanted to grant wage increases while keeping price stability, and saw higher productivity as the only way of doing both. The presence of a number of government agencies among the clients of P must also be taken as government support of automation. The same goes for BB, a public institution.

In India, the setting of X, unions are strong enough to be independent of both management and government. Here government practises an essentially conflict-mitigating role in response to automation. Official spokesmen seem to be on all sides on the pros and cons of electronic data processing and automation, but in practice management can apparently introduce it up to the point of causing excessive conflict.

* * *

The purpose of this summary of the case studies has been to highlight and pull together some of the diverse threads. The reader can check this evaluation for himself and identify other common or contrasting patterns that may have been omitted. No conclusions or policy implications are suggested here since these are set forth in the following section, "Summary and Policy Recommendations", which draws not only on the case studies but also on their consideration by all participants at the round-table discussion.

SUMMARY AND POLICY RECOMMENDATIONS

SUMMARY AND POLICY RECOMMENDATIONS

EFFICIENT RESOURCE ALLOCATION

Economic development means gaining higher standards of living through higher productivity. Waste and higher productivity are contradictory. Therefore no policy against unemployment can be against higher productivity or long-term efficiency. The task is to define these terms properly so that an optimum policy can be developed. That statement applies to many economic issues, but with respect to advanced technology and automation it gains a special poignancy. As Mr. Tévoédjrè, Assistant Director-General of the ILO, said in opening the round-table discussion: "Computers and manpower must not be on a collision path. Each must be deployed so as not to destroy the potential of the other." But the truth is that advanced technology *is* destroying the potential of much manpower in less developed countries. Not all such technology has this effect, and the exact proportion is debatable; but without doubt much of it substitutes without brilliance for manpower in simple, repetitive tasks. The case study of the Dacca computers showed that under-utilised computers were added to manpower that was already under-used. The computers were not used to bolster management so that new income- and employment-creating activities were generated. One of their main effects was the lowering of employment and income equality.

Macro-employment effects

The mere profitability of automation or computer installation does not prove that investment was efficient. It is easy to sell at a price higher than cost where, as is typical in poor countries, government tries to promote development by subsidising costs and by letting prices rise in protected and monopolistic markets. By simply looking at cases, therefore, one cannot tell whether or not something good, bad or indifferent is happening from the economy's point of view. In the cases reported at the round table, the companies rarely laid anyone off, because all of them were expanding. If these companies could profitably expand more, and if others could also expand, then unemployment would eventually disappear. The false implication is that unemployment is due to a lack of effective demand in poor countries.

What we have instead, however, is a problem of lack of complementary factors: insufficient capital. In their protected markets, are the advanced technology users wasting capital with their subsidised production techniques? In the words of Professor Ranis:

> Presumably firms, if they are properly maximising, will do what they think is best, and there I don't think we can be smarter than they. The question we can raise is, how accurate are the signals to which these firms are responding?

The implication of the question is that only better functioning of the market can help to prevent harmful uses of far advanced technology. Ranis insists that:

> Electronic data processing is not somehow a separate animal around which we must dance as around a golden calf. It's not separate from other technology. The same kind of criteria ought to apply to it as to other things.
>
> The method of intervention (to put it very crudely, and therefore inaccurately) can be either using direct controls and negotiating with thousands of decision makers on an ad hoc basis or setting some general rules . . . and then permitting the market as thus changed, if you like, to work.

If necessary, tariffs, subsidies and taxes could be used to offset other market imperfections and tendencies that are either hard to identify or difficult to attack directly. These measures may not even be sought by workers currently employed in an industry. Their true beneficiaries are the unemployed, the submarginal agricultural workers, and newcomers to the labour force. These are the groups that were slighted by the hothouse industrialisation policies during the 1940s and 1950s. Premature automation slights them today. Perhaps it is useful to have a national commission advising a government on automation, as India has today; but the best advice that can be given may simply be to let advanced technology come in on market criteria—that is, without special promotion or prohibition.

No one believes that the consequences will be easy to judge. An innovation will raise labour productivity physically, but then prices may fall as a result of the increased output. In time, however, people may shift their expenditure toward the more efficiently produced commodity, and a company's revenues would recover. If costs have fallen, profits will have risen, creating a potential pool for investment elsewhere. If these other activities have not been affected by similar innovations, average productivity there will be lower, thus bringing down the national average for employed workers. The high productivity investment will have been a smaller portion of the national total if these old activities also expand. Nevertheless, the country as a whole will fare better, since both larger employment and larger output will have been generated in the process. Note that much depended on whether prices were reduced and profits reinvested by the initially innovating company.

Foreign trade repercussions

In the last paragraph we mentioned productivity, profits and prices, but not wages. With conventional assumptions, economic theory would suggest that under conditions of surplus labour a small increase in employment and productivity should not affect wages. Where government and trade unions can invalidate the conventional assumptions, however, pressures can easily arise to let wages rise in proportion to productivity. Such a rise in wages would offset a number of the beneficial effects mentioned above, such as a possible fall in prices, the expansion of output and employment in accordance with demand elasticity, the rise in profits, and the consequent incentive and ability to invest. One important additional effect has not been mentioned: repercussions on foreign trade.

Where productivity rises without a concomitant rise in wages, a country has a chance of being internationally competitive—perhaps reducing imports, but, more important, perhaps raising exports. Many international time-series and cross-sectional analyses of economic growth have indicated that the growth of exports is one of the best, if not the best, predictor of economic growth as a whole. Conversely, stagnation in a country's export fortunes has been the most conspicuous handicap, at times paralysing countries that seemed to be attaining all the other prerequisites. Where wages in every branch are encouraged to rise with productivity (especially average, rather than marginal, productivity), a country may weaken some of its best pockets of comparative international advantage.

One might even go so far as to say that the effect of far advanced technology on exports is a crucial test of its wisdom. If exports are encouraged, that is sufficient to let a sector modernise. But note that this is not a "necessary" test. Some economists have been so appalled by the paralysing effects of balance-of-payments crises that raising exports or lowering imports seem to be the only legitimate goal of investment. Here the necessary condition is simply that the introduction of advanced technology has a higher marginal growth-stimulating effect than alternative ways of using resources. This may be the case where exports are unaffected. But where exports are raised, one can today hold as a crude rule of thumb that the evidence is sufficient.

On the difference between protecting foreign exchange reserves through higher exports rather than through lower imports, we can do no better than to quote Dr. Baranson:

> Make-work policies, through protection carried out indiscriminately and too far, are simply a short-term employment policy. Development and growth sooner or later catch up with you, and you have created the seeds of stagnation and inability to create future income and employment.
>
> Countries should take on selectively what they're capable of doing—not try to take on any imported component. They should be completely open and say: "You come in here, and if you export and earn foreign exchange, you can write your own ticket."
>
> In a restrictive economy they won't let you bring in the technicians. They won't let you bring in just the things you need. With openness you can make all the technical adjustments, and you can get a fit that lets you trade at 20 per cent below Japanese or United States prices.

A good example of such selectiveness in import substitution, cited by Baranson, was Argentina's policy toward computers. To balance the foreign exchange paid for imported computers the Argentinians contracted to make one piece of equipment (a card sorter) for the world market. Hence a little bit of foreign exchange is not saved on partial production of a wide variety of computers and the like; but much foreign exchange is earned on manufacturing, with modern large-scale efficient methods, a *single* piece of equipment.

In contrast with this, the large diesel engines described in Baranson's paper (and in the sources cited there) cost almost three times more when manufactured in India (US$ 6,500) than when manufactured in the United States (US$ 2,200). Because the wrong kinds of steel and the wrong methods of case hardening were used, it took several years simply to perfect components like filters and bolts. The former line of small Kirloskar diesel engines for irrigation pumps had been internationally competitive—20 per cent of the output had been exported. Indian economic policy should have built on that healthy base, if necessary allowing advanced technology to be an input, but not prematurely forcing it on large diesel engines.

Doubts about factor markets

The logic of efficient resource allocation through the discipline of the market is not easily contradicted. Where markets and market behaviour, for one reason or another, have been thoroughly instituted, basic advantages in resources and foreign trade will be reinforced and high growth rates will enable other problems to be solved as they arise.

But in circumstances of uncertainty, shortage, discontent and conflict, the discipline of the market, with its long-run criteria, is hard to accept.

For example, trade unions will demand to share in productivity increases that result from automation or computers. To maintain industrial peace, management and government will tend to agree. Disproportionately high wages necessarily call for disproportionate further modernisation. In one sense this is waste, but in another it is part of the price of obtaining any growth at all.

Professor Kassalow is an exponent of this kind of realism:

> I'm sceptical of the clearcut development models as applied to industrial relations questions. If a country is well managed and if employment opportunities are growing, unions may accept less than a full share of productivity increases. But the difficulty is—I'll exaggerate now—that Professor Ranis would confront the unions or the worker with sort of a last-stand neo-classical model of development.
>
> The countries don't accept this. Whether they should or not is another matter. Should the workers be content if they maintain their wage at some going level and trust that the benefits of advanced productivity will be distributed (presumably by the market economy) in the form of higher employment and expansion on the part of the firms that are earning the higher profits? Very few economies have developed this way. Wage differentials have been characteristic both of Western industrial nations and of a number of less developed countries who have made it in recent years, including Japan in the 1920s. Development is just not that smooth and even a process. If you don't have these kinds of safety valves of action that we would call trade unionism, you are likely to get other great interference with the whole modernisation process.

Dr. Routh goes even further—from realism about appeasing potential sources of trouble to scepticism about the efficiency of the most well-designed market economies (and Dr. Johri would agree):

> We should treat market signals with a good bit of suspicion because they have their origin in *a priori* economic reasoning. They may be wrong. The light may be shining green when it should really be shining red. They are more generally ignored, and they can be ignored for long periods without anything disastrous happening. And one never really knows whether one has rightly interpreted the signals until after the event. It's only *ex post facto* that you can see whether your interpretation was correct.

Specifically, one should be consistent in not making advanced technology subject to stringent market criteria when these are not applied elsewhere in the economy. If computers, for example, are under-utilised, that condition is hard to judge without knowing the degree of under-utilisation of other types of equipment in all sectors of the economy. According to Johri, under-utilisation in India was more or less equal in all industrial sectors.

In support of Kassalow's position, Johri cites figures of a growing disparity between Indian industrial wages and the incomes of rural and urban property owners. If the gross domestic product is growing and none of the increase accrues to workers, the likelihood of damaging industrial disputes is great.

The answer of Ranis to these observations is that:

> This is a picture of reality, yes, but not a favourable reality. No one here suggests that a *laissez-faire* nineteenth-century prescription would make sense. I'm not suggesting that the world is simple or can be simply modelled in a textbook sense. But I think you have to start with a framework. Positive resource analysis should come first.
>
> If income redistribution is necessary, keep wages low, and redistribute after production has been efficiently organised. And for that is advanced technology inevitable and desirable? Is this not unduly fatalistic making peace with reality—that is, an unnecessary reality?

In order to answer these questions, one needs to know more than the chances of conflict in a particular setting. Lack of other information is an impediment to the smooth functioning of markets in poor countries. Even worse is the uneven access to information. Professor Farouk stresses that:

> The salesman of a machine company could be more shrewd than the top man in the government or a firm in many of the developing countries. When an advanced technology like the computer comes in, a bank or a firm has almost to do nothing in the first years except to watch and listen to the supplier of the machinery and to dream of future profits.

On the other hand, Professor Kilby is sceptical of the vogue for second-hand machinery that was echoed in Ranis's paper. Kilby's experience in West Africa has been that:

> Whether they be private, indigenous or State enterprises, in a great many cases firms have failed when they have used second-hand equipment.

It takes tremendous technical knowledge and organisational skill to match second-hand machines into a working set. Paradoxically, foreign investors with the best access to capital can most readily exploit this sort of capital saving.

In fact, no one feels that a labour-intensive production function is always superior. Nor is anyone against flexibility. But productive flexibility depends on mitigating market imperfections as much as possible, both those related to social conflicts and those related to ignorance.

Dynamic effects

In considering the effect of either markets or market imperfections on the spread of advanced technology, one must see the efficiency of resource allocation in a dynamic rather than a static sense. "Dynamic" in this sense means that the aim is growth, not only of income but also of the factors of production that produce income, the quantity and quality of workers, capital and knowledge. Perfect market signals can lead to a perfect allocation of *given* resources, but they cannot convey anything about potentialities that are unknown to the buyers and sellers. They do not convey all the information needed for the best resource development to entrepreneurs because none of them is basing his own bids and offers on such knowledge.

These statements apply especially to the use of advanced technology from abroad. Markets channel some but not nearly enough knowledge internationally. To make up the deficiency, the Japanese sent hundreds of technicians abroad before the First World War. Some African countries, according to Kilby, send a few "showcase" entrepreneurs abroad over and over again because they make a favourable impression on foreign aid donors. The dynamic effects of such programmes are not nearly comparable to those of the Japanese. No wonder that the bids for advanced and alternative technology do not correspond to the actual potential of industrial branches!

In any case, all types of technology have output, employment and dynamic learning effects, with the latter perhaps being the most important and the most difficult to appraise. It is generally known that in developing countries, innovating "on top of" the imported capital-intensive technique is most likely. Ranis suggests in his paper that, in Japan, innovating on top of the imported technology in a strongly labour-intensive direction was accomplished on a large scale and that this possibility is underestimated in poor countries (Farouk does not share Ranis's optimism on this point). Of course, the kind of innovating involved here is less a matter of novel hardware than a matter of scaling down, of the conversion of skills, or rearrangements that are not spectacular. From an inexperienced economy's point of view, the simple production of spare parts can be a laboratory that teaches many things to its producers. From that they can go on to other things.

This is not to say that all dynamic effects are desirable. Undesirable things could be learned, and desirable skills forgotten. Foreign aid specifically could bring in techniques that are too advanced and that reinforce undesirable structural change. One cannot say that anything learned is *a priori* better than nothing. Johri, however, insists that one can also remember obsolete techniques too long. One should not be overwhelmed with apprehension, since advanced technology takes a greater lead time. If one always waits for an adequate demand, one may not have time to learn when market signals finally become favourable.

LABOUR RELATIONS AND ADJUSTMENT PROBLEMS

Problems of labour relations and adjustment to advanced technology are not distinct from questions of efficient resource allocation, static or dynamic. The type of settlement with labour can deflect the choice of technique toward a better or a worse alternative. The choice of technique can disturb labour relations since adjustment to some techniques is easier than to others. Thus some less developed countries are said to have labour legislation that is too advanced for the availability of factors of production. Mr. Schlesinger asserts that this is the case in Colombia. Among other effects, Colombian laws facilitate wage increases but make hiring and firing difficult. Here, obviously, is a market signal distortion that accelerates the spread of advanced technology.

But the aversion to wage differentials on account of institutional arrangements of one kind or another can be overdone. Much depends on how homogeneous the employed, underemployed and unemployed labour force actually is. Shifting workers from one category to another is a process not only with gains but also with costs. Johri observed:

> One of the purposes of wage differentials is to relocate the human resources in favour of technologically advanced and progressive industries. In less developed countries there should be greater wage dispersion than in advanced countries. This is a natural market phenomenon. If you tamper with this, there will be problems in the allocation of labour resources.
>
> Moreover, there ought to be some distributive process that gives the workers some share in productivity gains and a feeling of partnership in the growth process. If you do not have this, you have an industrial relations problem.

In these respects, one can possibly consider India as the vanguard of the less developed countries. Just as European countries follow the United States experience in many (but not all) economic patterns with a lag of a decade or so, perhaps less developed countries similarly follow India with a lag. Thus India already has 100 computers in commercial use with another 100 on order. Four companies have already been licensed to manufacture computers. These computers, of course, represent only one type of advanced technology.

Professor Dayal's position on industrial relations is close to but not identical with Johri's:

> Where unemployment is as much of a problem as it is in India, trade unions have no choice, in my view, but to make this a serious issue in any discussion on automation or mechanisation that reduces the total complement of labour.
>
> In India, whenever a textile mill has installed computers it has had no problem at all with the trade union. However, computers are used largely for decision purposes, for making more data available. This kind of development I see taking place more rapidly in India. Where it does not affect employment markedly, there would be considerable growth in this particular area.
>
> However, automation in manufacturing, numerical machine systems, high-speed equipment of one kind or another, which directly influence the manpower requirements of the industry, would have very slow growth.
>
> I would like to contrast this with newer factories, which are coming up in the chemical industry and in the electronics field. They are highly automated plants.

There will be increasing numbers of such technologies available in India in relation to new plants. When it comes to renovation of older technologies, when it comes to installing newer high-speed machinery or automated plants within the existing complex, I think that there would be a far more cautious approach and the process of change-over is likely to be very slow.

The largest number of automated plants in India is found in the chemical industry. Traditionally, relationships between employers and employees in chemicals have been good. Trade unions in chemicals are a part of the national network. All-India bodies have unions which work mainly in the chemical industry. I do not see any possibility of a national agreement between the chemical industry and the all-India body. I believe that with regional bodies agreements are possible. This chemical industry has its own problems. If it were to get mixed up with the national-level unions, I do not think that there would be any possibility of labour and management working out a satisfactory solution.

Wherever white-collar workers are involved in computerisation the resistance is far greater. In the case of blue-collar workers and automation, the response is correspondingly less.

This survey of India bears out one common theme in the case studies of this book. Advanced technology is usually introduced in such a way that redundancy does not arise among a firm's current workers. The only exception was Metal Box in Tanzania. In this case, according to Mr. Juma, the union had not only raised wages but also, when the occasion for dismissals came, eliminated arbitrariness and uncertainty. In the words of Routh: "Uncertainty is the root cause of industrial unrest."

Obviously no one is in favour of uncertainty *per se*, and no one nowadays is simply against unions *per se*. The question is, when do unions overreact to uncertainty and adopt measures that are harmful to the larger national interest? Kassalow has an answer to this question that is based on his own research and coincides with the banking case outlined in Dayal's paper. Kassalow says that unreasonableness and instability in union policy are due to the tensions that arise from a threat to a union's survival when that threat comes from a competing union. But even here the harm can be exaggerated. Referring specifically to rationalisation in Indian textiles, Kassalow says:

> One of the complicating factors of trade union action is that it tends to have a fairly limited time horizon. The truth of the matter is, however, that the leadership of the union did try to compromise, did recognise some of the difficulties and did attempt to reach some kind of agreement; and, in a pragmatic kind of way, rationalisation slowly went forward. Mass unemployment was avoided, and somehow the industry has not yet been overwhelmed by Japanese competition.

Kassalow recognises that when a firm is asked to carry extra workers to ease the transition to automation, profits for reinvestment elsewhere and productive employment of workers elsewhere are reduced. Dean Selassie not only recognises this effect but also asks whether it is right that progressive firms should pay the penalty of absorbing workers displaced by progress.

Johri adds that society may only think that it is transferring the cost of redundancy to firms. Competition does exist in some industries of poor countries, and under those circumstances the extra cost may lead to bankruptcy or failure to adopt an otherwise desirable innovation.

The trouble with all this, according to Ranis, is that objectives that should be separate are confused. Private firms should be harnessed to the development process and have the task of being as efficient as possible when the market signals are correct. Firms should let the chips of hardship fall as they may, because they should be able to let government be responsible for consequences that involve equity as opposed to efficiency. No one should suffer unduly; but no one should get premature welfare privileges. These cause premature automation and delay full employment. If government lets some workers gain special privileges, it should recognise that it ought to saddle itself, and not any special group of the private firms affected, with aggravated employment problems. Government should set income standards not for flourishing industries but as a residual public obligation for those who cannot find jobs. Then it will be clear if the standards conflict with employment.

Trade unions and government policy

Experience with industrial relations in developed countries for over a century has shown that the responsibilities of management, unions and government cannot be neatly separated and specialised. Employment transactions necessarily involve two parties, and an activist government always makes the groups involved a trio. Hence we have the tripartite tradition that extends to the ILO itself. Each group has its special objectives which involve trade-offs against the objectives of other groups. Moreover, the objectives of the other two groups, from the point of view of each of the three, may seem to be pure self-interest; but if they are neglected or impaired, the repercussions will inevitably undermine attainment of each group's own goals. If the various groups therefore participate in evolving one another's strategy, then misunderstandings, errors and conflicts will be fewer.

Few would disagree with this approach. To quote Ranis once more, a way of following it takes place when:

> ... unions are brought into the decision-making process by governments and are faced with this issue of the tremendous increase in the labour force that is threatening the less developed countries in the next 10 to 20 years and are asked to worry with the government decision makers about the disenfranchised, not yet employed workers, about those who are still in the agricultural sector, and about those who are not yet born into the labour force. The educational value of bringing unions into this kind of discussion would be enormous even if they are not going to sign on the dotted line.

Such an approach is good, but not as effective as it might be if the workers were able to draw on the same expert knowledge as management and government. The educational level, not only of the rank and file, but even of the leaders is likely to be much lower than that of spokesmen for management and government. This disparity is likely to be greater than in developed countries. Moreover, the financial resources will be less, so that outside consultants cannot be engaged to the same extent. Consequently, the ability to contribute and the ability to learn will be lower than one would hope for. Here, according to Routh, is a role for international expert assistance:

> I think it is very difficult for trade union officials, with their limited technical knowledge and also, possibly, with their limited social vision, to be able to understand the imputations of these technological transformations that are taking place. And I wonder whether an increasing role should not be taken by the ILO in this respect on behalf of the trade unions, the present labour force, and workers yet to come whose jobs (or lack of jobs) are going to be influenced by these techniques. It seems to me that a balance is not achieved in these cases, because typically you have a government agency or large employer, and possibly a firm of business consultants, and you have a computer salesman with his organisation behind him; and very telling arguments can be put forward for the installation of the computer, given certain instances. What we need is a highly informed agency that is able to give impartial advice on alternatives to computerisation . . . it does not seem to me that there is any agency capable of filling this particular role other than the ILO.

Kassalow suggested that the ILO might get tripartite approval for a code of responsibilities for management and labour—a kind of checklist of "good practice" with respect to labour legislation, collective agreements, dispute settlement and personnel policy. If such a code leads to agreement and co-operation, then it will be possible to bring in advanced technology, not only where it is rational but also with a rational process of adjustment for displaced workers. But in a disturbed social setting, a procedure that is perfect according to the code may aggravate tensions and conflict just as much as an imperfect approach. The probability of disruptive conflict varies from place to place and cannot be eliminated entirely by a fairly uniform checklist or formula. The seriousness of disruption or "flak" and its consequences in the short run and long run are hard to assess. But just because there is a wide range of possible settings, and of preferences within these, one need not conclude that newcomers to the problem of advanced technology and its manpower effects have nothing to learn from experience elsewhere. One can indicate both the limits of what is rational and a zone of what is most practical. Routh's view is that:

> The firm should not be expected to deal alone with this problem of redundancy. It is sometimes not possible for the firm. It should then be dealt with by the union in conjunction with employers in general. The position of the white-collar unions in Chile, cited by Professor Kassalow, which bargained through

a special agreement, provided that there would be no lay-offs among existing personnel. Natural attrition would be the source of any employment reduction. To the maximum extent possible, already employed workers were to be trained for the new computer positions, and employees who had to be shifted because of the impact of the computer would suffer no loss in pay. Employees who had been laid off would be rehired. This possibly may be a sort of model or prototype of agreement which we might think is reasonable and desirable for unions to try to negotiate.

Training managers for electronic data processing

A common view is that management may be impulsive or compulsive, instead of creative, in adopting computers. The opinion is that these deviations go beyond errors in adopting other types of equipment. Of course, such miscalculations apply not only to private management. Professor Strassmann reports that the public sector in the capital of one South American country paid over US$ 2 million annually to a computer-leasing foreign company in the late 1960s. By January 1970 the country had eight second-generation computers and seven third-generation computers. Most of these computers were used for only 160 to 180 hours per month, that is, less than eight hours per day for five days. And yet the agencies with second-generation computers were all ordering new third-generation computers with larger capacity.

One agency offered to share its new computer with another, but the latter feared that its priorities and confidentiality would be compromised. Spending money for excess capacity seemed preferable. This country's resources would have been better used for a large computer centre or two, tied in with smaller computers on the users' premises. The larger storage at the core would have allowed acceptance of programmes at more powerful terminals in the satellite centres. With co-operation and flexibility, a good computer manager would have no difficulty in scheduling priority jobs, large jobs, programme debugging, test runs and experimental work.

To this South American example, we can add Dayal's view that only one in six Indian computers is used effectively. Nor were they bought as a result of perceived necessity in the first place, as suggested by Johri's example of Company X. Where ignorance and impulsiveness is as great as this, perhaps a central team of government experts could act as an educational and screening device. After all, even in developed countries the rate of success according to expectations was only 30 per cent. A computer needs an elaborate preliminary study by skilled systems analysts, and even then an introduction period of four to five years is typical. Given these lags, it may be safer to introduce the latest model of computer that can fit a particular case.

Of course, circumstances can also exist where resistance to computers is excessive. For example, Kassalow cites French experience where middle management, because it was afraid of losing power, felt threatened by the computer. This resistance has even prevented the introduction of computers in some sectors where they might be productive.

According to Johri, in Company X the computer was viewed as disruptive by middle management for two years, and this view was self-realising. As a matter of fact, some parts of top management shared the view that disruption of normal working routines could not be tolerated. One might of course suppose that the computer had the very purpose of facilitating the shake-up in top management that took place, that it was the same sort of "magic wand" as a foreign adviser—an excuse to do things that were obvious all along. Hence a question arises, in the cases where the computer seemed to be a failure as a production tool: was it not a success as a management tool? At least in the case of Company X, according to Johri, such a shake-up was not part of the intention behind its introduction. In fact, the silk mill was already in trouble, and some kind of a change involving status positions in this authoritarian firm was needed in any case.

Professor Costa also insists that at Banco do Brasil SA no kind of management shake-up through the disruption caused by the computer was intended. At all stages efforts were made to avoid disruption, partly by training employees adequately and partly by going through the intermediate step of mechanisation. Computerisation was undertaken according to the most careful calculations. If a bank was far from a computer centre the break-even point would be 70,000 accounts (which corresponds to six accounting machines); but if the branch was close to a computer centre, the upper limit might be three accounting machines or 40,000 accounts.

In Baranson's view, the selective and stage-wise introduction of computers by Banco do Brasil SA was well handled. Nevertheless, according to Strassmann, one must remember that (estimating crudely) the monthly cost of computer use was equal to 150 jobs in Brazil at the equivalent of US$ 0.61 per hour—assuming the computer cost US$ 100 per hour. At the US$ 0.18 per hour wage for lower division clerks in Dacca, the equivalent employment lost would be 550 jobs.

Educational alternatives to advanced technology

Training is not only a complement to far advanced technology, as discussed in the previous section. Training can also be a substitute. One can substitute human capital for physical capital. This substitution has the advantage of substituting capital that is itself labour-intensively created for capital that is built with other physical capital or bought with scarce foreign exchange. In fact, the scarcity of capital and foreign exchange in poor countries is probably so great that the lag in incomes would seem hopeless if it had to be bridged with physical capital alone (even if that embodies principles of automation, electronics, computers and other modern technology). One must therefore examine where more training can substitute for more machines. In this section, we discuss that question—not generally, but in relation to more machines embodying the most advanced technology.

According to Schlesinger, the question is more complex than as just posed, because the training of managers to organise their firms better may lead to reductions in the firm's employment and to the use of more expensive equipment. Without looking at specific production methods, generalisations are difficult. In some cases, in Johri's view, building an efficient organisation may even be more expensive than buying computers if management and labour are really intractable.

Although he deplores the employment effects of computers, Farouk agrees with Johri on this point. With any conceivable management organisation in Dacca that would be purely labour-intensive, a certain point would come where the burden of inappropriate delivery records, charges and rebates would be so great, where backlogs would arise so often and be of such magnitude, that the incentives for raising efficiency would produce only minute results at astronomical cost. The tendency would be simply to give up.

Kilby agrees:

> People can be trained at reasonable cost to compete with electronic data processing in some, but not all, cases. One must do some things manually. One should not attempt to do everything manually.

With respect to substituting trained organisation and labour-intensive methods for automated production lines, Selassie notes the higher risks and suggests that it is simply safer economically to use automation. Who can assess how safe is safe enough? If introduction is to be selective, what is really being selected is the standard of safety. This is a matter that goes beyond equilibrium or shadow market prices.

In this connection, Kilby notes that to begin with labour-intensive second-hand equipment is the optimum choice for the national investor, especially when run by expatriates behind a high tariff wall, because it represents a smaller commitment. But once a market is ensured, interruptions of production become less tolerable, and new equipment will seem safer. Yet Kilby goes on to say that:

> Systems analysis without computers does exist (and indeed the ILO has a tradition of work in this field), and that is in the productivity demonstration area. In the mid-1950s the ILO did a great deal and got very astonishing results. Systems analysis at the plant level is industrial engineering, looking at plant lay-out, material flow, work methods, application of time and motion studies: scientific

management under its old name. The demonstration is very important because it has an electric effect when you show that a plant could, in fact, increase its output by 40 per cent without increasing any inputs. Foreman and supervisory courses are but part of a total package.

But optimism about the potential of training managers and improving organisation with conventional or labour-intensive technology is not unanimous. Baranson says:

> I think that there is a tendency to believe that, if only we had better-trained managers, things would be well in the world. We are dealing with a complexity that is much too great to say that. It takes 15,000 processes to make a diesel engine. It's just not the kind of thing (and I don't care how trained your managers are) that you digest in months or years. The job is just as difficult for the well-trained manager, the technology donor, as for his local counterpart, who is trying to absorb a whole set of routines that are unfamiliar to him.
>
> It's much more difficult to work with a scaled-down technique instead of using a piece of automated equipment. To take that and adapt it, to break it down into five or six machines, runs into the skill and manpower shortage which is a very formidable block, and it just isn't overcome in months or even years. It takes a long, long time, assuming that you have the conversion capabilities—but that's the other basic deficiency. Japan, after several generations of industrialisation, now has a very substantial supply of conversion skills, the people who are going to pick the other machines and draw the blueprints, and make all the adjustments. On a single part it's just fantastic to see the set of new processing sheets and the very careful detail. When you weld a special steel on to a part by a blow-torch instead of automated equipment, instructions must cover the size of the flame, the angle of the flame, the distance of the tip of the point, the test for the temperature of that flame, a colour test for the spread of the metal. Simple methods are much more difficult. The way you hold the blowtorch has to be exactly right if you're going to do it by hand. This is the great dilemma.

These difficulties are, moreover, reinforced by the restrictions put on international trade. Operating with only five qualities of steel, for example, is as difficult for any foreigner as for a native when the original design called for 20 types of steel.

Nevertheless, modern manufacturing can be made to work at international levels of efficiency in many branches, especially those where scale conditions have been met. Work can even be restructured to take advantage of local variations in factor prices. It is obvious in many countries that critical skills, and the education that can produce them, are lacking. Here a good strategy might be to begin by examining the employment, training and investment strategies of successful international corporations.

The role of culture

Many of the issues that have been discussed raise the question of culture. What kind of conflict and disturbance is likely in the culture? What kind of government policy toward technology or business will seem legitimate? What kind of training and education can management and labour absorb? Does advanced technology or labour-intensive training have a harder or easier time in non-Western cultures? Can the experience of Japan and the Republic of Korea be copied in Latin America, India or Africa?

In Kipling's day, East and West were thought to be wholly alien to each other. The more powerful West sent out missionaries and drill sergeants to change the mentality of the weaker East. Only when the Bible and the whip had done their work could trade and investment become productive. The economic sphere could not be isolated from others.

This view was rejected between the two world wars, when the West had to fight against racist dissension within. By the late 1940s productivity was to be raised in newly freed former colonies and other poor countries without questioning or interfering with their values. But output did not grow nearly as fast as dispensers of technical assistance and of long-term loans had hoped.

An intermediate position seemed to be indicated, and this applies as well to the issue of advanced technology. For example, Baranson says:

> I like to think that we have come full circle and that we recognise development as a sort of frontal effort where all aspects have concomitant ingredients. Yet the cultural affinity for industrialisation can be overdone. Labour after a few months can be disciplined and efficient. In Taiwan, girls take and cut quarter-sized wafers into 900 parts under a microscope, and then they weld these on gold wires with laser guns.
>
> This is where creative management comes in: accommodating workers according to culture in appropriate groups, restructuring the work, making it more or less repetitive. Routines can be adapted to life modes for more effective performance.

Ranis agrees that indispensable cultural affinity can be overdone, that *ex ante* people think the situation is different and are therefore unwilling to try remedies that worked elsewhere. Specifically, if the nineteenth-century Japanese techniques have been found to work outside Japan, why not give them a further try in India and Nigeria? The 200 years of nearly total isolation certainly helped to differentiate Japanese culture, and in their effort to industrialise the Japanese did not avoid all mistakes. But, Ranis continues:

> Because they permitted a greater play to the market as affected by over-all government direction, they were able to do it in 40 or 50 years. Most backward, feudalistic societies can change surprisingly rapidly with correct market signals.
>
> Government officials just do not have enough faith in the capacities of their own citizenry. There are all kinds of arguments made why the government has to do this and that. The government knows which are the priority industries; the government knows who should get the licences. I'm giving the government every benefit of the doubt that they have the best intentions. I just do not think they have the information on which to make these decisions and not enough fingers with which to implement them.

This position holds that governments and markets everywhere have the same potential and limitations, and that it is not "Japaneseness" *per se* that produced successful selectivity about advanced technology or anything else.

In Baranson's view, however, there was something special about Japan. He found that the Japanese engineer will constantly "hound" his foreign counterpart, asking for more and more information. His self-esteem is not eroded by asking question after question. He feels that he is exposing curiosity, not ignorance. In southern Asia Baranson found that the questioning stopped much earlier.

Dayal contests the validity of this comparison, but recognises important differences between Japan and India. According to him, Japanese industrialisation was so rapid not because culture does not matter where policies are correct, but because the Japanese culture is sharply different from culture both in the West and in countries like India. Duplicating the rigid stratification, understandings and tolerances of Japan would be impossible in India. The Japanese tend to delegate far less authority than either Western or southern Asian countries. Their authoritarian approach (and that of the Germans) has been less effective in India than British or American collaboration. When it comes to advanced technology, one should remember that the world has only 10 or 15 years of experience with things like computers. How can this be compared to the sweep of the entire nineteenth century? Nowadays all the minor conflicts, failures and inefficiencies of that era, which were very upsetting then, tend to be forgotten. Certainly the potential of second-hand equipment is not being any more ignored in India than it was in Japan. Ahmedabad itself has a flourishing market in used machinery. Moreover, India has attained the ability to export machine tools and sewing machines that are no longer being produced with French or Swiss technical assistance. Advanced technology is used where appropriate in their production. Chemicals and pharmaceuticals are being produced with automation and are also exported.

Kilby too sees cultural differences. He refers to such characteristics as responsibility, discipline, and willingness to co-operate on the job, which he contrasts with the touchy status-mindedness that interferes with production in a number of countries.

Tests of propositions involving culture come up against difficult "if"-questions. If the Japanese had not also engaged in imperial expansion throughout this period, would they have been better or worse off? Would advanced technology have had more or less of a place? If the Indians had limited the domestic content of those diesel engines to 10 or 15 per cent, would they still have cost three times the foreign price? If the United States had the Japanese ability to scale production methods downward, how much better a donor of advanced technology would it be?

In any case, it is interesting that opinions on culture and on the potency of better market signals did not coincide. Strong advocates of better factor pricing may either consider cultural differences important or unimportant variables . . . and vice versa. In other words, according to some experts, the nature of the technology and the productive setting outweigh culture as a determinant of the direction of policy. Given the direction, cultural differences can nevertheless explain variations in the speed of change. Others feel these purely culturally determined variations will be negligible.

POLICY RECOMMENDATIONS

The preceding discussion has been more than a preliminary setting of the stage for policy recommendations that have been deferred until this section. Policy has been at the centre of the analysis throughout. The first point to be discussed was the pure economic efficiency of advanced technology compared with alternatives and in relation to factor pricing. Policies towards technology and toward incomes and prices were examined. Foreign trade policy and long-run economic strategy were brought in. Later came discussions of policies toward trade unions, labour legislation and educational policy. Culture was discussed as either reinforcing one kind of policy or another, or as having negligible policy implications.

Nevertheless, it may be useful to spell out the policy implications once more, with all the analytical pros and cons of the earlier sections, but in a way that reflects the consensus of the round-table participants in July 1970. Since votes were not taken, the account is the judgement of the editors. (The gist of these policy recommendations has already been given at the end of the Foreword.)

1. Capital-intensive methods must not be subsidised when their main effect is to increase unemployment. Concern for the direct and indirect harm caused by unemployment is almost universal among experts in development. Capital-intensive methods may have advantages that outweigh the employment-creating effects of other methods, but when these advantages are small and doubtful it is better to err in the direction of maintaining employment rather than to adopt costly automation and computers prematurely.

Since adoption often depends on cost estimates, it is important that capital be not too cheap or labour too expensive, so that the investor's choice will coincide with the scarcities in the economy. Equipment must not be subsidised directly or indirectly. It must not be subject to unduly low tariffs, higher tax exemptions, easier credit or even subsidised servicing (although that is a complex matter). Wages of employed workers should not be increased without considering how this may influence the job opportunities of those affected by urban or rural unemployment, open or disguised. Trade unions must share responsibility for the healthy functioning of the entire economy, and they should therefore participate in the making of a national incomes and employment policy. The ILO can help train officials for this task.

2. Shortages of skilled managers must be corrected by means of training and the provision of an economic setting that encourages learning through experience. Such shortages can be partly overcome through the use of automatic machines and computers since these are often more easily managed than people. But that is an expensive capital-using approach. Training managers, on the other hand, is labour-intensive. Managers must be trained to lay out better plants and offices, to schedule production more efficiently within these, to set up more manual quality control systems, and in general to motivate subordinates in a larger hierarchy more effectively. Some of these things can be taught in schools; the rest comes with experience reinforced financially. Small failures must mean a financial penalty that is noticeable but not crippling. Small successes must bring a reward that is gratifying but not satiating. Where markets do not provide this kind of elastic response, they may be bolstered by government action to just that extent. Unfortunately a danger exists that government will raise small stimulating market pricks into numbing or satiating injections.

3. Advanced technology should never be simply forbidden on the grounds of employment repercussions. Where the productivity and profitability of ultra-modern equipment overcomes the properly weighted private and social cost of all inputs, it should be seen as a strength, not a weakness. Productive equipment usually means high profits, high savings and low prices. In some cases, it represents an opportunity to penetrate foreign markets and to earn foreign exchange. In others, local productive effort will be spurred by the availability of better rewards in the form of cheaper consumer goods. At the same time, the higher savings can make more capital available to other industries than seemed originally to have been diverted from them to pay for the advanced technological installation.

4. Policies concerning computers should fit the above recommendations. Computers should be correctly priced in terms of the true scarcity of capital and foreign exchange. They should not be used for routine internal data processing just because present management is incompetent to handle a sufficiently large staff. They should not simply be forbidden.

Large, complicated operations do have a place in some industrial, commercial and public sectors of poor economies, and some of these can benefit from a continuous flow integrated information system that only computers can provide. The difficulty is that possible gains are hard for management to judge and easy for computer salesmen to exaggerate. An advisory service by government or perhaps by the ILO could be useful here.

In most cases, industrial problems and equipment are too heterogeneous for intelligent appraisal by outsiders. But computers are specifically used for that minority of problems that different firms and industries share, that is, problems at a generalised level of accounting. One does not have to understand the physical characteristics of materials and products, and the process that converts one into the other, in order to assess their financial characteristics. By the same token, a computer can be shared by firms engaged in greatly differing activities. Since neither sufficient prior appraisal nor reasonable sharing of computer time after installation has been characteristic of computer diffusion in poor countries, both should be strongly encouraged through energetic training and co-ordination policies. Agencies for the purpose must be established. ILO help seems possible.

5. Workers displaced as a result of the introduction of advanced technology should not be penalised. The responsibilities of management, workers, unions and government in cases of technological redundancy should be codified in a manner that is both fair and consistent with a rational manpower adjustment policy. A guide, checklist, recommendation or other measure could be developed under the auspices of the ILO.

6. Continued research into the problems and patterns of technological change is usually recommended by researchers, such as those who attended the round table. This round table was no exception. Special stress should be laid on the alert analysis of successful but partly accidental solutions of problems. What will it take to transfer these solutions to other settings? The more that is known not only about the settings but also about the effects of different kinds of advanced technology, the smoother and more correctly paced will introduction be. How do different types of advanced technology vary in their effect on skill and job content? Are the psychological and sociological effects of office automation in poor countries the same as those in developed countries, and as in other industries? Can jobs be redesigned, especially supervisory jobs, in order to modify the expensive spread of highly advanced technology? What training programmes are best? What have been the most successful cases of redesigning jobs, products and technology for the conditions of developing countries—in other words, how can the benefits of advanced technology be secured without penalties arising from the fact

that this technology was created for a different environment? General answers to these questions may be less useful than those aimed at a specific industrial branch. All these studies could be sponsored or carried out by the ILO alone or jointly with other United Nations agencies.

* * *

One matter that both policies and research should stress in relation to advanced technology is the time horizon. Differences in recommendations or criteria often reflect differences in the time horizon that has been chosen, consciously or merely by implication. Bringing implicit choices into the open tends to encourage more rational policies. Moreover, a long time horizon also tends to encourage greater consistency among the policies of different groups, whereas a short time horizon tends to make labour oppose advanced technology where it is needed, while making management support it where it is not needed. Labour's fear of losing jobs can preserve inferior techniques and higher costs that impair export competitiveness and therefore employment. By contrast, management with a short time horizon will not want to bother to rethink and to scale down productive processes and to retrain a staff of intermediate supervisors. Automating is easier, though perhaps more expensive in the long run. Myopia tends to polarise reactions and to increase the chance of conflict, needless conflict. If this book can serve as a lens with some corrective long-range focus, the hopes of all the authors will have been more or less realised.

APPENDIX : LIST OF PARTICIPANTS

Dr. Jack Baranson, International Bank for Reconstruction and Development, 1818 H. Street, Washington, DC.

Professor Celso Albano Costa, Banco do Brasil SA, DESED, Av. Rio Branco 120-9º, Rio de Janeiro.

Professor Ishwar Dayal, Indian Institute of Management, Vastrapur, Ahmedabad 15.

Professor A. Farouk, Head of the Department of Commerce, Dacca University, Dacca 2.

Dr. C. K. Johri, Associate Director, Shri Ram Centre for Industrial Relations, 5 Pusa Road, New Delhi 5.

Mr. Omari S. Juma, Friendship Textile Mill Ltd., PO Box 20842, Dar es Salaam.

Professor Nicholas Kaldor [1], King's College, University of Cambridge, Cambridge.

Professor Everett Kassalow, Department of Economics, University of Wisconsin, Madison, Wis.

Professor Peter Kilby, College of Social Studies, Wesleyan University, Middletown, Conn.

Professor Gustav Ranis, Department of Economics, Economic Growth Center, Yale University, Box 1987, Yale Station, New Haven, Conn.

Dr. Guy Routh, School of African and Asian Studies, University of Sussex, Brighton, Sussex.

Ato Seyoum Selassie, Dean, School of Social Work, Haile Sellassie I University, PO Box 1176, Addis Ababa.

---

[1] Professor Kaldor presented a paper to the round table but was unable to attend.

Mr. Daniel Schlesinger, University of the Andes, Calle 97 13-14, Bogotá.

Professor W. Paul Strassmann, Department of Economics, Michigan State University, East Lansing, Mich.

INTERNATIONAL ORGANISATIONS

Mr. Crespin, Organisation for Economic Co-operation and Development, Development Centre, 91 boulevard Exelmans, Paris 16e.

Mr. Mir Khan, Consultant to the Administrator, United Nations Development Programme, Palais des Nations, Geneva.

Mr. F. Muller, Economic Affairs Officer, Industrial Division, Economic Commission for Europe, Palais des Nations, Geneva.

INTERNATIONAL LABOUR OFFICE

Mr. Albert Tévoédjrè, Assistant Director-General.

Mr. R. H. Bergmann, Chief, Automation Unit.

Mr. J. Schregle, Chief, Labour Law and Labour Relations Branch.

Mr. C. Hsieh, Economic Branch.

Mr. M. Kabaj, Economic Branch.

Mr. V. V. Lukin, Management Development Branch.

Mr. K. Marsden, Small-Scale Industry Section.

Mr. C. O'Herlihy, Economic Branch.

Mr. L. Pisarik, Vocational Training Branch.

International Institute for Labour Studies

Mr. R. Ray.